Robotics and AI:
Building Practical Intelligent Robots

机器人与人工智能：
智能机器人的实践与开发

Amir Ali Mokhtarzadeh(艾米尔·A. 莫赫塔尔扎德) ［英］

高尚兵　肖绍章　周　泓　孙成富　编著

东南大学出版社
SOUTHEAST UNIVERSITY PRESS
·南京·

图书在版编目(CIP)数据

机器人与人工智能：智能机器人的实践与开发／(英)艾米尔·A.莫赫塔尔扎德(Amir Ali Mokhtarzadeh)等编著. — 南京：东南大学出版社，2023.9
 ISBN 978-7-5766-0753-6

Ⅰ.①机… Ⅱ.①艾… Ⅲ.①智能机器人-研究 Ⅳ.①TP242.6

中国国家版本馆 CIP 数据核字(2023)第 087972 号

责任编辑：刘　坚(635353748@qq.com)　责任校对：周　菊
封面设计：王　玥　责任印制：周荣虎

机器人与人工智能：智能机器人的实践与开发
Jiqiren Yu Rengong Zhineng: Zhineng Jiqiren De Shijian Yu Kaifa

编　　著	Amir Ali Mokhtarzadeh　高尚兵　肖绍章　周泓　孙成富
出版发行	东南大学出版社
社　　址	南京市四牌楼 2 号　邮编：210096
经　　销	全国各地新华书店
印　　刷	广东虎彩云印刷有限公司
开　　本	787mm×1092mm　1/16
印　　张	16.25
字　　数	420 千字
版　　次	2023 年 9 月第 1 版
印　　次	2023 年 9 月第 1 次印刷
书　　号	ISBN 978-7-5766-0753-6
定　　价	68.00 元

本社图书若有印装质量问题，请直接与营销部调换。电话(传真)：025-83791830

About the Author

Amir A. Mokhtarzadeh, a British male born in February 1961, is a professor at the Huaiyin Institute of Technology, School of Computer and Software Engineering, Robotics and AI Labs, and a master supervisor. Amir graduated from Manchester Metropolitan University (UK). Until 2000, Amir was a computer programmer and senior manager for the industry, with many achievements and accreditations to complement the projects he led. From 2000, when he started teaching in higher education, he had responsibilities as a researcher in funded projects, a senior lecturer, and a divisional leader.

He is passionate about teaching and leading young researchers toward developing innovative and creative projects. He began developing a robotics research lab at HYIT's "Computer Science and Software Engineering" faculty in 2015. His lab was visited frequently by local officials and continuously recognised as a "foreign expert studio" by the Huai'an municipal government and Jiangsu provincial government. He won the Friendship Award of Jiangsu in 2019 and was awarded the Envoy of the People's Friendship of Jiangsu in 2022.

As a senior lecturer for the last twenty-one years, he taught many IT and computer-related subjects in multi-disciplinary courses such as Linux, Java, ROS, deep learning, and image processing. In 2003, he was awarded for attending a dinner reception at No. 10 Downing Street, London (invited by Prime Minister Tony Blair) together with some managers and lecturers in the UK.

Amir actively presents Artificial Intelligence (AI) and Robotics related subjects at various conferences and universities, such as in 2016 at the International Conference of Robotics (Huaiyin Institute of Technology, Huai'an, Jiangsu), in 2018 at the Deep Learning Industry Conference (2018 Energy and Power Engineering International Cooperation and Exchange Conference, Changsha University of Science and Technology, Changsha, Hunan), and in 2018 at the Impact of Artificial Intelligence Conference (2018, IEEE, Energy and Power Engineering International, Shenyang, Liaoning).

Introduction

This book aims to bridge the gap between industry and academia. Despite the brilliant performances of many excellent graduate students, they need to demonstrate their practical knowledge and ability in interviews. Starting Industry 4.0 required completely different technologies in production and preparing students to participate in such a fast-developing world of automation and robotics. According to Statista, the global robot market is expected to grow by around 26% by 2025. As a senior lecturer for the last twenty-one years, I realize that this rate will necessitate a rapid development of practical knowledge and experience.

When I was studying for my Electrical and Electronic Engineering degree in the early 1980s, I learned how to build my first microcontroller board using the Zilog Z80. Creating a microcontroller-based project back then required board-level development and machine language programming skills, and I used that board to develop a simple text-to-speech project. Nowadays, with the vast amount of resources available to our students, they are expected to be engaged in state-of-the-art and mind-blowing AI projects. They can contribute to future generations if we lead them on the right track and enrich their multi-dimensional skills.

Among so many published and online resources, what do I expect to deliver through this book? Well, with the current tools and resource kits around pre-designed hardware and pre-defined software for robotic labs, there are minimums that a robotics lab in higher education can offer to increase creativity in students. Academics must catch up to industry and avoid losing sight of their purpose and goal at the start of the fourth stage of the industrial revolution. Using bare development boards with bare hands requires educators who can handle hardware. Of course, prebuilt hardware is easy, convenient, and modular for teaching in a robotics lab, both for organisations and educators, but certainly not for students. Students will attempt to join online groups, communities, and resources with their ever-growing enthusiasm and learning goals. Similarly, while offering a hands-on approach to robotics projects, this book is also trying to be a valuable companion for the theoretical side of related projects. Therefore, this book can be considered a companion to many robotics courses by creating a link between the classroom and the robotics lab.

Today, despite many online servers providing AI services, such as face recognition or voice recognition, they cannot offer us a mobile robot to deliver those AI methods or help you build an arm with a precise trajectory. Still, we will need to learn how to use SLAM (simultaneous localisation and mapping) to design an autonomous robot or use inverse kinematics to move a robot arm to a precise position.

Every section of each unit ends with problem-solving questions, exercises, and experiments for robotics lab activities. Each unit ends with an introduction to a practical project for students to develop.

This book is an extraction of studies, research, and lectures between 2011 and 2022. During all those times, I always had two questions in my mind:

One: How can I make computer programming more exciting and meaningful?

Two: How can I encourage innovation and creativity in an age of resource sharing?

Amir A. Mokhtarzadeh

Acknowledgements

While this book will discuss robots and projects built at the Huaiyin Institute of Technology during 2015−2022 specifically, in general, its main focus is a deep dive into the practical making of robots as well as programming. Through this journey, we will introduce many open-source pieces of hardware and some open-source software, as well as non-open-source devices. With such a wide area of discussion, it is difficult to know how deep to dive into each subject to fulfil the purpose without straying too far from the scope of this work.

The author would like to thank the following students for their contributions to this book:

Gu Xiaoshi, Fei Hongyan, and Du Meng.

Also, students who made contributions to the development of Robotics and AI Labs since 2015:

Sun Shiyao, Cheng Yujian, Ai Ming, Zhao Zhangwei, Xiao Long, Ming Hui, Abu BakorHayat Arnob, Kabu Chimdia Primus, Hudoyberdi Toshboev, Gao Han, Kong Rongshuang, Ding Nan, Ge Junzuo, Li Biao, and many more.

CONTENTS

Chapter 1 The First steps 001

    What Is a Robot? 003

    What Is Robotics? 003

    Robots and AI 004

    Robotics Lab 006

    Research Paradigm 006

    Hardware and Software in Robotics 009

    Basic Electronics 009

    Components 012

    Software Packages Used in Robotics 020

    Linux in Robotics 021

    Digital and Analog Signals 025

    Digital Signal Processing 026

    Number Systems 027

    Logic and Binary Systems 034

  Exercises and Projects 037

Chapter 2 Introduction to Embedded Robotics
.... 039

  Microcontrollers and Development Boards 041

    MCU and Development Board 041

    Development Boards and Power Consumption 076

    SBC (Single Board Computer) 078

Sensors 089
Sensor's Classifications 090
Transfer Function & Sensor Characteristics 095
Sensor Modules 099
Sensor Shields 100
Optical Sensors 101
Temperature Sensors 109
Proxy 115
Motors 128
DC Motors 128
Servo Motors 129
Stepper Motors 130
Hardware Control 132
PID Control 132
PWM 137
Controlling Servo Motors 140
Manipulators-I 142
Advance Use of Microcontrollers 147
Dealing with 'Delay' 147
Multitasking and OO Programming 152
Object Oriented Programming with Microcontrollers 152
User Interfaces 157
Exercises and Projects 159

Chapter 3 Robotics Design and Control 161
Different Robots 163
Tools for Intelligent Robotics Design 164
Image Processing 164
ROS (Robot Operating System) 170

Manipulators- II ··· 177
    Calculating Forward Kinematics ··· 179
    Ground Moving Robots ··· 185
Toward Intelligent Robotics ··· 201
    Intelligent Vision Sensors ·· 201
Exercises and Projects ·· 209

Chapter 4   Artificial Intelligence and Machine Learning ············ 211

    Introduction ·· 212
    Terminology ··· 213
    What Is Neural Network? ·· 214
    AI Ready Devices ·· 233
Human Intelligence vs. Robot Intelligence ····························· 237
    How Robots See the World ··· 238
    Dispositive Networks and Understanding of the World ············· 239
    Human-Robot Interaction ·· 240
Exercises and Projects ·· 242

Chapter References ·· 243

Chapter 1

The First Steps

Chapter 1 The First Steps

What Is a Robot?

Contrary to popular belief, getting experts to agree to answer this question is extremely difficult. For many, a robot is a mechanical machine to perform physical tasks; for others, it is more related to decision-making. The reason may be that the concept of robots is more dynamic and interactive, which can evolve with our knowledge and expectations.

Robots are artificial agents that function in the real world. The goal of objective robots is to manipulate items by sensing, picking up, moving, altering their physical characteristics, destroying them, or having an effect on them, freeing up human labour from monotonous, distracting, or exhausting repetitive tasks.

Defining a robot
- Must be able to move
- Must be intelligent
- Must have power
- Must be able to sense

What Is Robotics?

Robotics is an artificial intelligence domain that focuses on developing effective and intelligent robots. A subfield of artificial intelligence called robotics uses computer science, electrical engineering, and mechanical engineering to design, build, and operate robots.

Robotics is a field of study that deals with programming physical robots to perform automated or semi-automated tasks. The tasks carried out by robots depend on our demands and expectations. At the third stage of the Industrial Revolution, our needs for such tasks were semi-automated to increase productivity. However, by the end of the third industrial stage, we desired fully automated tasks to be carried out by robots to improve precision and replace the workforce. Therefore, all concepts in the robotics field have been built around the mechanical and automation concepts of tasks performed by robots. In academia, mechanical engineering and automation faculty were leading the study of robotics. Computer programming is mainly used for control and operation.

During the last two decades, the robotics field has improved substantially due to many factors, including the Internet, open electronics, GPUs, increased storage capacity and prices, and the availability of a large number of data sets.

By development more

Social robots

Self-driving cars

Voice assistance systems

Use of robots in medics

Now many other fields are attached to robotics, such as medical science, social science, philosophy, and more. Developing intelligent robots is possible with the help of those fields of science. For a self-driving car to predict a pedestrian's intention, integration of mutual understanding between robots and humans is necessary. Social robots must be able to make sense of human actions and intent to perform tasks in our space.

Robots and AI

For many, there is no clear distinction between robotics and AI, which becomes more apparent when we consider the stages of the industrial revolution.

Phase 1: using mechanical machines running on steam power

Phase 2: Steam power is replaced by electrical power. In addition, the invention of electricity created a new era for electromechanical devices.

Phase 3: The invention of CPUs opens the road to microcontrollers and programable control. This phase is known as automation. During this period, automated machines such as manipulators replaced humans in many production processes. The word "robotics" was used instead of "automation," like "robotic arm," "robotic control," etc.

Phase 4: With improvements to machine learning and deep learning frameworks, new forms of automation have started that don't follow a flow of programme sequences but can make decisions based on trained data. The robotic concept in a more proper form has received attention from many institutions and industries. Also, it needs to be mentioned that AI is not robotics.

Table 1-1 Comparison Between AI Programs and Robots

AI Programs	Robots
The input to an AI program is in symbols and rules	The input to robots is analogue signals
need general purpose computers to operate on	need special hardware to operate in different fields, to perform various tasks
usually operate in computer-stimulated world	operate in real physical world

As AI becomes more popular, many products use AI names. We must have a clear understanding of AI and robots.

Nowadays, in higher education institutes, computer labs give their places to robotic labs. However, this transition is slow, and sometimes, because of a lack of resources, it needs to be in the right direction. Industry 4.0 requires new talents, innovations, and creativity. Traditional automation methods are inadequate for us to be on the cutting edge of the emerging industrial revolution-stage 4. Automation was mainly based on repetition with sensory decision-making, while new robotic fields require intelligent reformulation and learning methods.

Robotics Lab

> Tell me, and I will forget,
> Show me, and I may remember,
> Involve me, and I will understand.
>
> —Confucius

For many computer science students, sitting behind a computer, writing a program, and running it on screen is typical but not very intuitive. Nowadays, most computer applications are developed by large corporations to provide cutting-edge graphics and functionality. Today, writing a programme with limited functionality by a student cannot contribute to a student's ambition or creativity. Instead, dynamic programming can be more exciting and encourage us to step up our ambitions.

Dynamic programming is when we can see the execution of our programme outside of a computer monitor, such as on a moving robotics arm or a task for a service robot. Many factors, including open electronic and internet standards, have made robotics projects more accessible in the last decade.

Robotics is not only about what we teach and what we learn. It's about how we do it. How do we take the theories for a walk? For computer software engineering and computer science students, Robotics Lab is a programming playground where they can see how their written codes can interact with the real world. They can create actions that can perform in the real world.

Research Paradigm

Starting a robotics project has some essential prerequisites.

First, being a good team player is essential to a successful project. It can guarantee a robotics project can reach its target. Recent robotics projects involving robot building have become more intensive, requiring more skill and time than a single person can manage. As a result, being a good team player is one of the essential prerequisites for success. That means being a good team player and:

- Understanding Open Source and understanding professional responsibility
- Sharing knowledge as well as achievements
- Asking questions
- Helping others

Also, it's essential on a personal level to be neat, tidy, and organised:

Good project management requires the following:

—Managing research and development

—A good time management

—Understanding priorities

—Track progress and recording

—Producing a periodic report

What Can a Good Robotics Lab Offer?

Robotics Lab is the first point of learning and development for our projects. A robotics lab full of resources is a gold mine. A robotics lab with only pre-built robotics parts and instructions to connect them in a specific order and execute prescribed programs, on the other hand, is a killer of creativity.

What a robotics lab must provide is a series of tools and devices that we can use to develop an innovative idea. If our projects use a range of hardware tools, software resources, and microcontrollers/selection of sensors, then we need to recall and review some electronic skills.

First, we need to know about tools and safety procedures. A robotics lab may provide tools such as a soldering iron, electric drill, or jigsaw. As a result, understanding the risk of using these tools by students and providing induction and training are an integral part of any project in a robotics lab.

Therefore, to conduct a project in a robotics lab, students may need to recall their introductory physics and electronic knowledge.

Safety Procedures

Some certain steps and procedures should be followed to maintain safety in a robotics lab. In general, the safety of a robotics lab in higher education depends on the structure of the lab itself. A robotics lab that only offers pre-designed small robotic modules with plug-and-play capability may require a very different safety protocol than a lab developing a large-scale robotics arm, humanoid robots, or quadruped robots. In general, we must consider three areas of safety before students start using the lab.

Safety Protocols for Using Tools and Electric Devices

Personal safety and precautions for working with live wiring and connections. Students must be well-trained in working with live electric wires and devices.

Safety of robotic devices. Robotic devices such as sensors and development boards can be costly. Having appropriate induction before using such devices saves students from failure and costs to the organisations.

Safety to others. Ensure neither the tool nor the robot could cause injury or danger to others.

Students should give special attention to the following general safety procedures in Robotics Labs:

- Keep the work area tidy.
- Wear protective gloves, glasses, and so on, depending on the tool.
- Do not leave a running device unattended. You must pause or turn off the device, even if you leave it for a short time.
- Wearing lab clothes or uniforms is safer than having loose clothes or hanging items like jewellery.
- Avoid drinking or consuming food in the work area.
- Always seek assistance from your instructor if you need clarification on using a tool or device.
- If you must use an extension cord, ensure it is not running through an unsafe or unprotected area.

- While touching digital electronic devices or boards, use an earthing wristband to protect sensitive devices from static electricity.
- Turn on heating or cutting tools while you are using them.

Despite many safety rules common to many labs, each robotics lab must produce safety procedures that depend on tools, parts, and activities specific to that lab.

Hardware and Software in Robotics

A robotics project always consists of hardware and software. Without using hardware, it is just a software project, and without software, it is considered a mechanical project.

Depending on the kind of robotics project, a certain level of electrical and electronic understanding is essential. Here, this section will discuss the most relevant topics in electronics. For some students in higher education courses, this section is considered a recall of previous knowledge; for others, it might be a chance to learn about new subjects.

Although students' reflections in higher education are expected to fit into scientific or engineering representations of theories and methodologies, I always choose simple and visual representations of theories for practical learning. Furthermore, as previously stated, the robotics field attracts learners from other disciplines who may require a refresher on basic concepts in the simplest way possible.

Basic Electronics

Terminologies and Basic Concepts

Before discussing any electronic components or methods, two very basic concepts must be made clear: one is voltage, and the other is current.

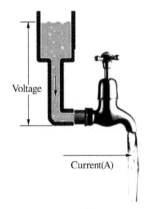

Figure 1-1 Current

Voltage (V)

If you were to think of electricity as a water tank, volts would be how high the water rises in the tank or the depth of water in the tank. Voltage is measured in volts (V).

Ohm Law

$$V = IR$$

Where V stands for voltage (in volts), I for current (in amperes) and R for resistance (in ohms).

Kirchhoff's Voltage Law

Kirchhoff's voltage law (KVL) states that the algebraic sum of all voltages around a closed path (or loop) is zero.

$$\sum_{m=1}^{M} v_n = 0$$

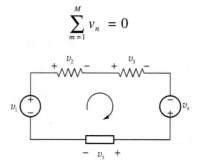

Figure 1-2　Kirchhoff's Voltage Law

Current (i)

If volt is the water depth in a tank, then current is the force with which the water is moving out of the attached pipe. Current is measured in amperes (or, colloquially, "amps").

Kirchhoff's Current Law

The total of all currents in each branch of a circuit is equal to zero.

$$\sum_{m=1}^{M} i_n = 0$$

Resistors

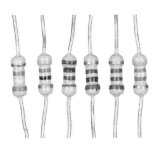

Figure 1-3 Resistors

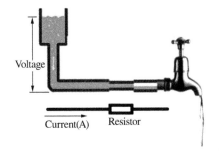

Figure 1-4 Effect of Resistance on Current

With the above example, the resistor would be the tap, which can limit the flow of water out of the tank. A resistor is a device to reduce the flow of electric current.

Any material has a certain degree of resistance to the flow of electric current. The flow of current is linearly proportional to the voltage across the unit. This behaviour is encapsulated in "Ohm's Law."

Resistors can be configured in series or parallel. One can add up the resistance; the other can divide it proportionally.

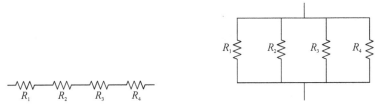

(a) Resistors connected in series   (b) Resistors connected in parallel

Figure 1-5 Resistors in Series and Parallel Connections

Resistors in series simply add their resistance:

$$R_{tot} = R_1 + R_2 + \ldots + R_n$$

Resistors in parallel add reciprocally:

$$1/R_{tot} = 1/R_1 + 1/R_2 + \ldots + 1/R_n$$

Components

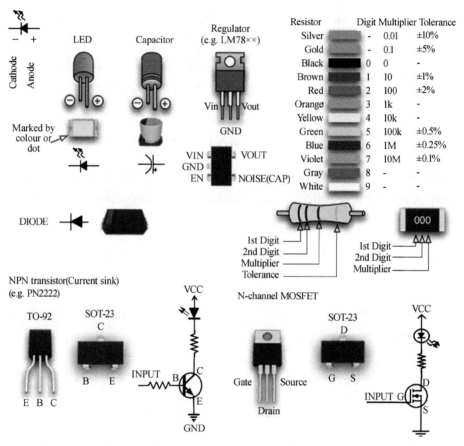

Figure 1-6 Common Electronic Components and Resistors Colour Codes[1]

Capacitors

Capacitors are used to store electrical energy through charge separation. A capacitor is composed of a pair of conducting plates separated by a thin insulator. When a voltage is applied across the plates, one becomes positively charged and the other becomes negatively charged.

Figure 1-7 Capacitors

Chapter 1 The First Steps

Capacitance, denoted by C, is measured in farads.

Definition for farads: One volt across one farad produces one coulomb of charge separation.

$$Q = CV \Rightarrow I = CdV/dt \Rightarrow V = \int I dt / C$$

Capacitors in series:

$$\frac{1}{C_{tot}} = \frac{1}{C_1} + \frac{1}{C_2} + \ldots + \frac{1}{C_{N-1}} + \frac{1}{C_N}$$

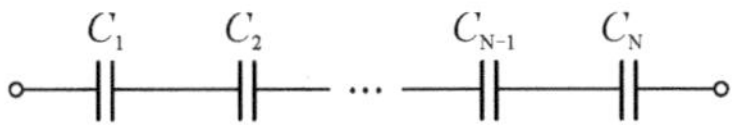

Figure 1-8 Capacitors in Series Connection

Capacitors in parallel:

$$C_{tot} = C_1 + C_2 + \ldots + C_{N-1} + C_N$$

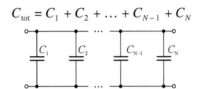

Figure 1-9 Capacitors in Parallel Connection

Batteries

Batteries are containers to store electricity at a set voltage. Their value is measured in amps. A higher amp value provides electric current in a higher amount or for a longer time. Batteries may be connected in series or parallel to increase their set voltage or electric current.

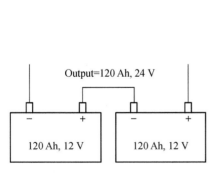

Figure 1-10 Batteries in Series

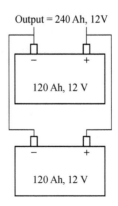

Figure 1-11 Batteries in Parallel

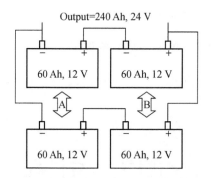

Figure 1-12　Batteries in Parallel and Series

LED (Light Emitting Diode)

LED is one of the most common components in any project because it allows for the simplest user interaction. LEDs can provide visual information at a very low cost and are easy to implement in every project.

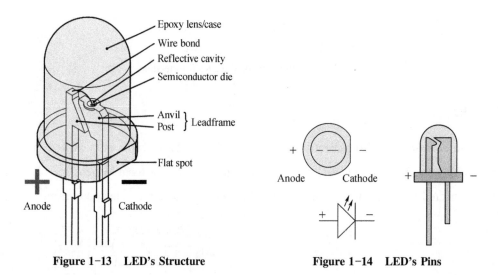

Figure 1-13　LED's Structure　　　　Figure 1-14　LED's Pins

LED diodes have two pins (anode and cathode). The anode is connected to the positive supply, and the cathode is connected to the negative supply. LEDs are extremely low-power components. To connect a LED, a resistor must be used to reduce the power to a safe level. Otherwise, it may be harmed or lose its life.

LED manufacturers determine the operating voltage and current. To calculate the resistor value, we can use the formula below:

Resistor = (Battery Voltage − LED voltage)/desired LED current

$$R = \frac{V_s - V_l}{I_l}$$

For example, assuming a 12-volt power source and a white LED with the desired current of 10 mA, the formula becomes: resistor = (12 − 3.4)/0.010, which is 860 ohms. Of course, resistor values calculated using the above formula may not be standard resistor values. In these cases, we can use the closest standard resistor. In the above example, the closest standard resistor is 820 ohms.

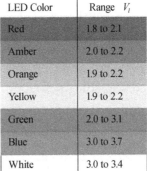

LED Color	Range V_l
Red	1.8 to 2.1
Amber	2.0 to 2.2
Orange	1.9 to 2.2
Yellow	1.9 to 2.2
Green	2.0 to 3.1
Blue	3.0 to 3.7
White	3.0 to 3.4

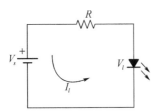

Figure 1−15　LED's Circuit　　　　Figure 1−16　LED's Colour and Voltage

Also, different LED colours have different voltage (V_l).

Tools

There are certain tools to be used for minimal electronic work, and sometimes it is impossible to do a task without them.

Digital Multimeter: to measure voltage (DC/AC), current, resistor, capacitor, and more based on the type.

Soldering Iron is essential to make connections more reliable and permanent than using connector wires.

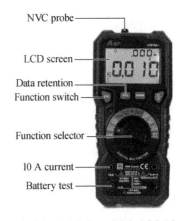

Figure 1−17　A Modern Digital Multi-meter

Figure 1-18　Soldering Iron and Accessories

Wires are another essential part of developing a
robotics project. Their range, aside from size, is single core, multicore, shielded, etc. Choosing the right wire depends on many factors, including the connecting device, functionality, and supporting power.

A good practise is to keep the colour code the same throughout a project. Most standard approaches are for red as positive supply (+ V, Vcc, +), and black as negative supply (GND, - V, -). However, when dealing with a live wire and connections (110 v or 240 v), we must strictly follow the standard colour code.

Figure 1-19　Connector Wires　　　　　　Figure 1-20　Wires

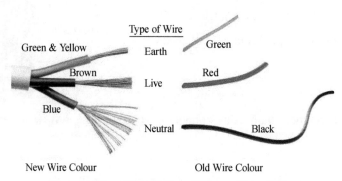

Figure 1-21　Old and New Colour of Live Wires

Chapter 1 The First Steps

For more information on basic electronic components and tools, there are many online resources available: https://www.instructables.com/Basic-Electronics-Skills-for-Robotics/.

Voltage Divider and Kirchhoff's Voltage Law

Imagine you have a power supply (a battery, for example) and want to reduce its voltage for your application. A voltage divider can divide the supply voltage into two different voltages by using two different resistors.

$$V_s = V_{R1} + V_{R2}$$

V_s: the supply voltage

V_{R1}: voltage across R_1 resistor

V_{R2}: voltage across R_2 resistor

A voltage divider divides the supply voltage. It is useful when we have a higher supply voltage than what we need.

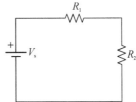

Figure 1-22 Kirchhoff's Voltage Law and Voltage Divider

If both resistors have the same value, then the voltage will be half at the output (V_{out}):

Since the current is the same in both resistors, the voltage is divided between the two; thus, it is a voltage divider. Voltage divider circuits are very common, even when one or both circuit elements aren't resistors.

Since $I_1 = I_2$, (by Kirchhoff's current law),

and $V_s = V_{R1} + V_{R2}$, (by Kirchhoff's voltage law),

so $V_s = I_1 R_1 + I_1 R_2 = I_1 (R_1 + R_2)$

$$I = \frac{V_s}{R_1 + R_2}$$

If R_1 gets smaller, then

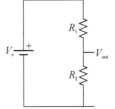

Figure 1-23 Voltage Divider Circuits

$$V_{out} = V_s \frac{R_2}{R_1 + R_2}$$

gets bigger.

If R_1 gets bigger, then

$$V_{out} = V_s \frac{R_2}{R_1 + R_2}$$

gets smaller.

As R_1 gets bigger, then

$$V_{out} = \left(\frac{R_2}{R_1 + R_2}\right)$$

gets smaller.

If $R_1 = 5 \ \Omega$, $R_2 = 10 \ \Omega$, $V_s = 5$ V,

$$V_{out} = \left(\frac{R_2}{R_1 + R_2}\right) = 5\left(\frac{10}{5 + 10}\right)$$

$$V_{out} = 3.3 \text{ V}$$

Calculating power consumption of a robotics project.

One of the very first tasks in a robotics project is calculating total power consumption. For a service robot or where a fixed power supply is not available, choosing the right size of battery can play an important role in our project. Inadequate power can lead to insufficient power and the failure of many parts or reliability. For instance, if we do not provide adequate power for a servo motor considering its maximum load, it may not behave properly, or a drop in power for a sensor may result in wrong data. To calculate power consumption, we need to know three terms:

- Current (amps)
- Voltage (volts)
- Electrical power (watts)

To calculate power draw, they're applied to a simple formula:

$$\text{Current (amps)} \times \text{Voltage (volts)} = \text{Power (watts)}$$

For example, with one of our projects we might list the following pieces of equipment:

- 1 × LCD monitor
- 1 × Servo Motor

Chapter 1 The First Steps

- 1 × SBC (Single Board Computer)

After looking up the manufacturer specifications, we find:

- LCD monitor has a 250-watt rated power supply.
- Servo motor can draw a max of 5 amps at 12 V.
- SBC has a 150-watt rated power supply.

To determine the maximum power usage at any one time, we calculate:

$$250 \text{ watts} + (5 \text{ amps} \times 12 \text{ volts}) + 150 \text{ watts} = 460 \text{ watts}$$

The maximum power usage for these three pieces of equipment is 460 watts.

Usually with robotics projects, we are more interested in maximum current consumption than power consumption to calculate the correct size of a power supply or a battery.

Current consumption:

The amount of current required by a device to operate is referred to as its "current consumption." Basically, every electrical or electronic part we use in our project has a minimum and maximum current consumption. This can be found by referring to the part's specification or directly calculated by using a multi-meter.

Suppose we have a project consisting of an Arduino UNO development board and three LEDs. An Arduino UNO on its own demands about 45 mA at 16 MHz. Each LED may use 15 mA. In total, all the parts taken together are:

$$45 \text{ mA/h} + (3 \times 15) \text{ mA/h} = 90 \text{ mA}$$

Your power supply must be able to provide your project with 90 mA to operate. However, in the case of a battery that has a limited source of power, its operating time depends on the battery's capacity. For instance, a battery with a capacity of 360 mA/h in your project can operate for four hours.

$$\frac{360 \text{ mA/h}}{90 \text{ mA/h}} = 4 \text{ h}$$

Software Packages Used in Robotics

Although there is no set rule or prescription for the software used in robotics labs and research, there are some advantages to using certain software over others due to its ease of use or greater resources.

Programming Language for Robotics

It is undeniable that Java and C++ are the most commonly used languages in every industry. They are both object-oriented programming languages with extremely strong reliability. However, when it comes to robotics, it is slightly different. It is not about production only; it's about resources, hardware, libraries, and mainly prototyping.

There is no specific programming language we can claim as a robotics programming language. Certain languages and libraries, according to reputation, provide more resources and facilitate development.

The two most resourceful languages commonly used in robotics are "C++" and "Python." What made Python suitable to be placed next to a powerful object-oriented programming language like C++ is its powerful resources in machine learning and its role in developing ROS (Robot Operating System) packages.

In recent years, the Python programming language has proven its power both in robotics and AI. There are many online resources available based on Python for nearly every piece of hardware and project.

With the IT revolution, and especially with the emergence of data science in recent years, the importance of Python has increased manifold since it has become the primary language in data science.

Despite Python's popularity in the robotics field, it is sometimes not a matter of choice and must be used. When programming at a low level, such as with microcontrollers or drivers, C or C++ is required. When we need to just develop faster and focus more on the project itself than the programming language, then Python can be a good choice. In addition, there are times Python and C++ get together to perform solid work when we use the Python binding.

C++ is a compiled language. Every time you make changes, you must compile and test your code, where Python is a scripting language and doesn't need to be compiled. Python allows for the rapid development of prototypes. With Python, you can quickly set up a script to test specific parts of your robotics coding. On the other hand, when you develop a project commercially and need licencing instead of developing open source, you will need to use C++.

Python is a very simple language with a very straightforward syntax. There are two major Python versions: Python 2 and Python 3. Python 2 and Python 3 are quite different. For instance, one difference between Python 2 and Python 3 is the print statement. In Python 2, the "print" statement is not a function, and therefore it is invoked without parentheses. However, in Python 3, it is a function and must be invoked with parentheses.

Different resources and devices may require Python 2. This tutorial uses Python 3 because it is more semantically correct and supports newer features. Then developing a project in a virtual environment is probably the best option.

There are many online python interpreters can be used for learning python fast:
https://www.tutorialspoint.com/execute_python_online.php
https://www.onlinegdb.com/online_python_compiler
https://www.online-python.com/
https://www.programiz.com/python-programming/online-compiler/

Also a 'Python 3' Cheat Sheet is provided at this book's online companion.

Linux in Robotics

In the recent decade, Linux in general and Ubuntu Linux in particular have played an important role in successful research and development in robotics labs. Perhaps the most important reason is its open-source licensing, stability, security, flexibility, and close relationship with Python, ROS, and OpenCV.

- Over 25% of professional developers use Linux-based operating systems. (Stack Overflow)
- Linux powers 39.2% of websites whose operating system is known. (W3Techs)
- Linux powers 85% of smartphones. (Hayden James)
- Linux, the third most popular desktop OS, has a market share of 2.68%. (Statista)
- The size of the Linux market worldwide will reach $15.64 billion by 2027. (Fortune Business Insights)
- The world's top 500 fastest supercomputers all run on Linux.
- Linux doesn't power only two out of the top 25 websites in the world. (ZDNet)
- Today, there are over 600 active Linux distros.

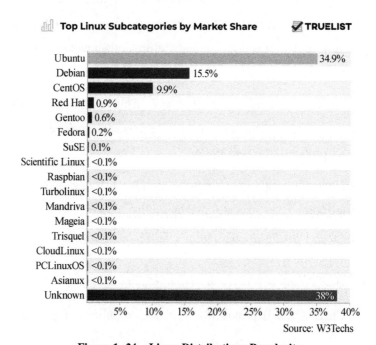

Figure 1-24　Linux Distributions Popularity

Starting with a bit of history can help us understand more about Linux.

All modern operating systems had their roots in 1969 when Dennis Ritchie and Ken Thompson developed the C language and the Unix operating system at AT&T Bell Labs.

The C language was specially developed for creating the UNIX system. It was much easier to develop an operating system that could run on many different types of hardware using this new technique. By 1975, when AT&T started selling Unix commercially, about half of the source code had been written by others. Other developers were disappointed as the company would sell their hardworking software developments for profit. Therefore a legal battle resulted in the existence of two versions of Unix: the official AT&T Unix and the free BSD Unix.

In the eighties, many companies started developing their own Unix:

- IBM created AIX.
- Sun made SunOS (later Solaris).
- HP developed HP-UX.

About a dozen other companies did the same.

The result was a mess of Unix dialects and a dozen different ways to do the same thing. And here is the first true root of Linux, when Richard Stallman founded the GNU project to end the era of Unix separation and everyone reinventing the wheel.

GNU is a Unix-like operating system. It is a collection of many programs: applications, libraries, developer tools, and even games. The development of GNU, which started in January 1984, is known as the GNU Project. Many of the programmes in GNU are released under the auspices of the GNU Project; those are what we call GNU packages. "GNU" is a recursive acronym for "GNU's Not Unix." "GNU" is pronounced g'noo as a single syllable, similar to saying "grew" but with an n instead of an r.

The nineties started with Linus Torvalds, a Swedish-speaking Finnish student, buying a 386 computer and writing a brand-new kernel called Linux Kernel.

In English, it means the central or most important part of something. In computer science, a kernel is the central part of an operating system. It manages the tasks of the computer and the hardware, most notably memory and CPU time.

In summary:

Linux is not an operating system. It is the **kernel**. GNU/Linux is an operating system. However, here we use "Linux" to refer to GNU/Linux.

A Linux **distribution** is a collection of (usually open source) software on top of the Linux kernel, which is called the GNU/Linux operating system.

A distribution, or "distro" for short, can put together server software, tools for managing the system, documentation, and many desktop apps in a central, secure software repository.

Figure 1-25   Linux Kernel

The Linux command anatomy is:

- $ *command* [option(s)][argument(s)]

The command identifies the command you want Linux to execute. The name of a Linux command almost always consists of lowercase letters and digits. Remember that, unlike Microsoft Windows, Linux is case-sensitive; be sure to type each character of a command in the proper case.

Below is a very limited list of Linux commands with only a few options as a reference.

Table 1-2

	Command	Description
Essential	man	To find information regarding certain commands
	cat/etc/os-release	To display the version details of the operating system
	head/etc/passwd	To display 10 lines from top of user related information
	tail/etc/passwd	To display 10 lines from button of user related information
	Ip addr/ifconfig	To display IP of the device
	Sudo apt-get	To install or remove new software packages. Options are as (install, purge, …). Example: sudo apt-get install…
	source	The source command can be used to load any functions file into the current shell

Chapter 1 The First Steps

(Continued)

		Working with directories	Options
Working with directories	cd	Change directory	.. ~ -
	ls	List directory contents	-a -l -h
	pwd	Print working directory	
	mkdir	To make a new directory	-p
	rmdir	To remove a given directory	-p
		Working with files	Options
Working with files	file	The file utility determines the file type	
	touch	To create an empty file	-t
	rm	To remove a file. Also, rm -rf can remove anything	-i -rf
	cp	To copy a file or directory	-i -r
	mv	To move a file or directory as well as to rename	
		Working with file contents	Options
Working with file contents	head	The head command can also display the first n lines of a file	-n -C
	tail	Similar to head, the tail command will display the last ten lines of a file	-n
	echo	Very strong command with the basic function as Input a line of text and display it	-e
	cat	To read file contents and concatenate files	
	More	The more command is useful for displaying files that take up more than one screen	
	less	Same as more	

Digital and Analog Signals

What is a signal? A signal is an electromagnetic or electrical current that is used for carrying data from one system or network to another. A robot's perception of the outside world is based on data received from its sensors. Sensors are used to capture information from the natural world as we know it. The majority of that information is in the form of analog signals. These kinds of signals work with physical values and natural phenomena such as earthquake, frequency, volcano, speed of wind, weight, and lighting.

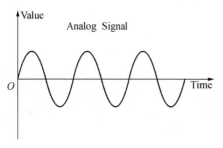

Figure 1-26   Analog Signals   Figure 1-27   Digital Signals

A digital signal, on the other hand, is a representation of only two states. While analog signals are time-varying, digital signals are time-separated signals. Digital signals have only two states: high or low. A digital signal can also be represented as a 1 or a 0.

Digital Signal Processing

Signals must be processed to extract the information they contain to be analysed, visualised, or even converted into another type of signal. The real-world information in the analogue form will go through an analogue-to-digital (A-D) conversion for further processing. Digital Signal Processing (DSP) is manipulating signals (such as voice, audio, or temperature) that have been digitalised to manipulate them mathematically. In other words, DSP performs essential mathematical functions on the digitalised data.

Our sensors are developed harmoniously with the natural world and can receive signals in analogue form. Although our computers require that we digitalise signals for processing, unlike computers, we can understand analogue data better. Therefore, DSP may then prepare the digitised information for use in the real world. It does this in one of two ways: digitally or in analogue format by going through a digital-to-analogue (D/A) converter. All of this occurs at very high speed.

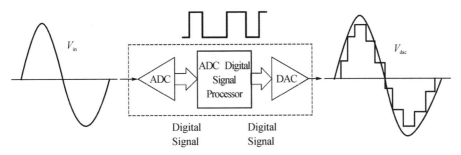

Figure 1-28 Digital Analog Signal Processing

There is no robotics project that doesn't need to use DSP. In fact, DSP is used in nearly every project where sensing real-world data is required. These include areas of the audio signal, speech processing, RADAR, seismology, audio, SONAR, voice recognition, and some financial signals. One of such examples is digital signal processing, which is used for speech compression in mobile phones as well as speech transmission in mobile phones.

Number Systems

We will always come across a programming task that requires a binary or hexadecimal conversion, but we will not always remember how to do it. This section is a review of the number system and the conversions between them.

Base-10 (Decimal)

When we count, we usually use Base-10 number system. This is a number system we are comfortable and our skills build around it. We use $(0, 1, 2, 3, 4, 5, 6, 7, 8, 9)$.

$1 = 10^0 = 1$

$10 = 10^1 = 1 \times 10$

$100 = 10^2 = 10 \times 10$

$1000 = 10^3 = 10 \times 10 \times 10$

$10000 = 10^4 = 10 \times 10 \times 10 \times 10$

...

...

To visualize how we understand a number in Base-10 consider number 231:

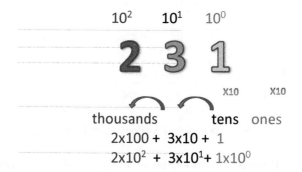

We can learn different ways to think about numbers or visualise them in our minds, but mathematically, it all comes down to the combination of 10 ×.

However, Base-10 is not the only number system around. There are other number systems such as Base-2, Base-8, and Base-16.

Base-8 (Octal)

Base-8 number system is called Octal. Base-8 means there are only 8 digits (0, 1, 2, 3, 4, 5, 6, 7) in this system. There is no such number as 8, or 9. When we count in this system 0, 1, 2, 3, 4, 5, 6, 7, then as in Base-10 system we continue with 10, 11, 12, 13, 14, 15, 16, 17, 20, 21, 22, 23, …

Then number 10 in Base-8 system represents 8 in Base-10 system, and 11 represents 9, and so on.

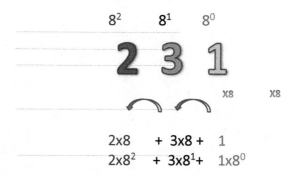

Chapter 1 The First Steps

Converting from Base-8 to Base-10

Question: What is 324_8 in Base-10 system?

If we want to convert a Base-8 number like 324 to a Base-10 number, we follow the procedure below:

$$
\begin{aligned}
&\text{Starting from right (ones):} \\
&4 \times 8^0 \quad\quad = 4 \\
&2 \times 8^1 \quad\quad = 16 \\
&3 \times 8^2 \quad\quad = 192 \\
&\overline{} \quad\quad \overline{} \\
&\quad\quad\quad\quad\quad = 212_{10}
\end{aligned}
$$

Converting from Base-10 to Base-8

If we want to convert 150_{10} to Base-8, first we find the largest power of 8 that is smaller than our number. Here, this is 8^2 or 64 (because 8^3 is 512 which is larger than our number). So, our first digit is 8^2.

$$
\begin{aligned}
&150/8^2 \quad\quad = 2 \text{ remainder } 22 \\
&22/8^1 \quad\quad = 2 \text{ remainder } 6 \\
&6/8^0 \quad\quad = 6 \\
&\overline{} \quad\quad \overline{} \\
&\quad\quad\quad\quad = 226_8
\end{aligned}
$$

A General Conversion Rules

To create a universal conversion method from any number system to the Base-10 system, we can build on the strategy we used to convert from Base-8 to Base-10.

If we have a number as $y_4 y_3 y_2 y_1 y_0$, where (y) is a digit in a number and 'c' is the base number value. Base-10 value would be:

$$\text{Base-10} = (y_4 \cdot c^4) + (y_3 \cdot c^3) + (y_2 \cdot c^2) + (y_1 \cdot c^1) + (y_0 \cdot c^0)$$

Example:

32311_4 conversions to Base-10 is:

$$3 \times 4^4 + 2 \times 4^3 + 3 \times 4^2 + 1 \times 4^1 + 1 \times 4^0 = 949_{10}$$

Base-16 (Hexadecimal)

Base-16 or hexadecimal is a number system with 16 digits as (0, 1, 2, 3, 4, 5, 6, 7, 8, 9, A, B, C, D, E, F). In this way A represents 11 in Base-16 system, B for 12, C for 13, D for 14, E for 15, and F for 16.

Because they make converting between the Binary and Hexadecimal number systems simpler, hexadecimal numerals are frequently employed in computer programming. Hexadecimal numbers, for instance, correspond to four bits per digit (binary digits).

Using the general rule for converting hexadecimal number to decimal is the same as with Base-8 numbers. Consider $A2B_{16}$ number:

Dec	Hex	Oct	Bin
0	0	000	0000
1	1	001	0001
2	2	002	0010
3	3	003	0011
4	4	004	0100
5	5	005	0101
6	6	006	0110
7	7	007	0111
8	8	010	1000
9	9	011	1001
10	A	012	1010
11	B	013	1011
12	C	014	1100
13	D	015	1101
14	E	016	1110
15	F	017	1111

Given hexadecimal number = $A2B_{16}$

Figure 1-29 Number Systems

First, convert the given hexadecimal to the equivalent decimal number.

$$\begin{aligned} A2B_{16} &= (A \times 16^2) + (2 \times 16^1) + (B \times 16^0) \\ &= (A \times 256) + (2 \times 16) + (B \times 1) \\ &= (10 \times 256) + 32 + 11 \\ &= 2560 + 43 \\ &= 2603 (\text{decimal number}) \end{aligned}$$

Base-2 (Binary)

There are only two digits in binary system, 0 and 1, HIGH and LOW, TRUE and FALSE. As computer systems all are based on binary system, learning binary system is so important.

Chapter 1   The First Steps

Table below will show the Base-2 (Binary) system.

Using previous methods to convert a binary number to decimal:

$$101100_2 = 1 \times 2^5 + 1 \times 2^3 + 1 \times 2^2 = 32 + 8 + 4 = 44_{10}$$

Similarly, converting a decimal number such as 13 to Binary would be:

Step 1: Divide the given number (13) repeatedly by 2 until you get '0' as the quotient

$13/2 = 6$ (remainder 1)

$6/2 = 3$ (remainder 0)

$3/2 = 1$ (remainder 1)

$1/2 = 0$ (remainder 1)

Step 2: Write remainders in the reverse order (1101)

Then $13_{10} = 1101_2$

Converting from Binary to Decimal:

Table 1-3   Decimal-Binary Conversion Table

Position	9	8	7	6	5	4	3	2	1	0
Power of two	2^9	2^8	2^7	2^6	2^5	2^4	2^3	2^2	2^1	2^0
Value	512	256	128	64	32	16	8	4	2	1
Binary	0	0	0	0	0	0	1	1	0	1

For example, from the table above, the decimal value of 13 is a sum of $8 + 4 + 1$, which can easily be converted to binary by adding 1's.

Conversion Between Binary and Hexadecimal or Octal

There is an easy way that will let you convert between binary and hexadecimal quickly. First, take any binary number and divide its digits into groups of four. So, say we have the number 1101011100_2. 0011 0101 1100 is the result of dividing the binary number into groups of four. Notice how we can just add extra zeros to the front of the first group to make even groups of 4. We now find the value for each group as if it were its own separate number, which gives us 3, 5, and 12. Finally, we simply use the corresponding hexadecimal digits to write out the Base-16 number, $35C_{16}$.

Table 1-4 Binary to Hexadecimal Conversion

Steps	Example
Start with a binary number	1101011100
Start with the binary number into groups of 3	0011 0101 1100
Convert each binary group into an octal digit	3 5 C
Combine your digits to get the base 8 number	$35C_{16}$

We can go the other direction also, by converting each hexadecimal digit into four binary digits. Try converting $B7_{16}$ to binary. You should get 10110111_2.

This trick works because 16 is a power of 2. What this means is that we use similar trick for Base-8, which is also a power of 2:

Table 1-5 Binary to Octal Conversion

Steps	Example
Start with a binary number	1101011100
Start with the binary number into groups of 3	001 101 011 100
Convert each binary group into an octal digit	1 5 3 4
Combine your digits to get the base 8 number	1534_8

Numeral System Conversion Table

Table 1-6 Numeral System Conversion Table

Decimal Base-10	Binary Base-2	Octal Base-8	Hexadecimal Base-16
0	0	0	0
1	1	1	1
2	10	2	2
3	11	3	3
4	100	4	4
5	101	5	5

(Continued)

Decimal Base-10	Binary Base-2	Octal Base-8	Hexadecimal Base-16
6	110	6	6
7	111	7	7
8	1000	10	8
9	1001	11	9
10	1010	12	A
11	1011	13	B
12	1100	14	C
13	1101	15	D
14	1110	16	E
15	1111	17	F
16	10000	20	10
17	10001	21	11
18	10010	22	12
19	10011	23	13
20	10100	24	14
21	10101	25	15
22	10110	26	16
23	10111	27	17
24	11000	30	18
25	11001	31	19
26	11010	32	1A
27	11011	33	1B
28	11100	34	1C
29	11101	35	1D
30	11110	36	1E
31	11111	37	1F
32	100000	40	20

Negative Numbers

For negative numbers, we adopt a new representation: that is, we use an extra bit (by convention, the most significant bit) to denote if a number is negative. In fact, we go one step further and express any negative number in the 2's complement form.

The 2's Complement of a Number

The following procedure is used to obtain a number's second compliment:

Starting from the least significant digit, and copy bit-by-bit till we reach the first "1". The first "1" (the least significant "1") is copied. All the remaining digits are inverted (a "0" is converted to "1", and vice versa.)

Example:

Binary number, b = 0010110

2's Complement of b = 1101010

Table 1-7 Number's Second Complement

Number(D)	Binary (B)	-2's Compliment (-n)
3	0:0011	1:1101
5	0:0101	1:1011
7	0:0111	1:1001

Logic and Binary Systems

The binary system consists of two states: digital signal and logical operation. Binary logic has only three operations that it can perform. These three binary logical operators (and, or, not) are powerful and form the basic design of computer systems. **AND** (indicated by the symbol "&" or "") means that, given two statements X and Y, if and only if both are true, then $X \cdot Y = 1$.

If either one is false,

then $X \cdot Y = 0$.

OR (indicated by the symbol ‖ or +) means that, given two statements X and Y, if at least one is true,

then $X + Y = 1$.

If both are false,

then $X + Y = 0$.

NOT (indicated by the symbol X') and operates on a single variable, if X is true,

then $X' = 0$

If X is False,

then $X' = 1$

The above logic operands and their truth table represented in digital electronic with the following diagrams as logic gates in digital electronic:

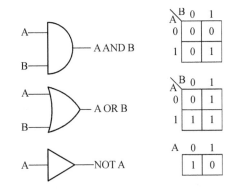

Figure 1-30 Logic Gates (AND, OR, NOT)

These three operands are unimaginable powerful in expressing logic statement and specially to build if-then expression[2].

For instance, consider the following example:

If it is rainy tomorrow, or it is cold, but not lover than 15 ℃, then we will play basketball.

This sentence can be represented with a variable X and three statements as:

X   we will play basketball

A   it is rainy tomorrow

B   it is cold

C   not lower than 15 ℃

$$X = A + B \times C$$

Using logic gates to draw the above example, we may do it wrong if just simply read it from left:

The above example may not be correct without using a correct order of priority of operands:

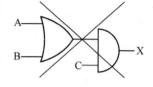

a. all NOT operators must be evaluated first;

b. all AND operators (starting from the right if there is more than one) second;

c. all OR operators (starting from the right if there is more than one) are evaluated third.

Then the correct form of the design is:

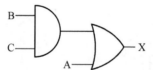

Figure 1-31   Priority Order of Logic Gates Operands

With combination of discussed three logic gates, there are another two very useful logic gates[3]:

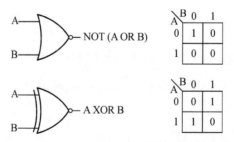

Figure 1-32   Logic Gates, NOT, and XOR

Using login gates, we may be able to design a more complex logic system. A good understanding of the logic and binary systems can be beneficial in writing programmes, especially in C++, as well as working with hardware.

Exercises and Projects

Changes in the field of robotics are speedy. Therefore, to provide more up-to-date problem-solving techniques and innovative projects, this part will be available from the author's website at www.roboticacamp.com/robotics_and_ai.

Exercises and Projects

Chapter 2
Introduction to Embedded Robotics

A system which contains a dedicated software application for a specific purpose and is embedded into a hardware design is known as an embedded system[4]. The centrepiece of all robots is an embedded system, which usually consists of one microcontroller and (or) an SBC (single-board computer). While this may be adequate for many small educational mobile robots, it may need much more for recent AI-powered robots with either onboard or remote GPU-powered machine learning algorithms and many individual microcontrollers interconnected to form a more complex control system.

Chapter 2　Introduction to Embedded Robotics

Microcontrollers and Development Boards

Microcontrollers have undergone major developments, and now nearly every electronic device uses them. Especially during the recent decade, various MCU development boards moved into education for prototyping. Some because of the development of Open Electronic in recent years, like Arduino, and some because of access to the power of single-board computers, like Raspberry Pi, which enriches students' experiences in all disciplines. Because of such developments, as well as many others, which are not limited only to IoT (Internet of Things), robotics, and cloud computing. Embedded systems are evolving into something greater than just controlling an isolated system. Ambient intelligence and ubiquitous computing are the outcomes of orchestrating those many parallel developments.

Here, the challenge, instead of looking at the structure of microcontrollers or their functions, as there are so many good books to cover such information, is to investigate the strengths of popular MCUs in a single study for their applications to answer the question about their use cases in such fast and exponential development, particularly in the field of intelligent robotics.

Many times, my students have asked me, "Are real robots developed using ARM or Atmel microcontrollers?" Although this question is completely wrong, there is still a point here for further discussion and exploration.

MCU and Development Board

There is always a question for students: what is the difference between CPU, MPU, and MCU?

When the first central processing unit (CPU) was developed as a single chip in the early

1970s, it was referred to as a microprocessor (μPs) and later as a microprocessing unit (MPU).

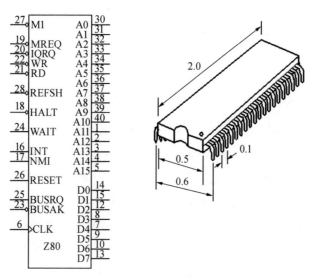

Figure 2-1   Zailog Z80 Pinouts

One of the early MPUs was the Zailog Z80, which I used in my project in 1984. In comparison to now, it was a difficult task back then, and we had to use many other components, such as a clock frequency generator, memory to complete my project.

We can have the opportunity to only concentrate on our main project using microcontrollers without worrying about peripherals, memory, oscillators, etc.

Then here, an interesting question arises: What is the difference between MPU (Microprocessor Unit), MCU (Microcontroller Unit), and microcontroller development boards or development boards? To answer this question, we must look inside MPUs and MCUs in a little more detail. Below is the block diagram of Z80 MPU.

A typical microprocessor is a chip with several pins which must be connected to input/output buses (Data and Address) and usually cannot do anything without additional circuits and components.

In contrast, a typical microcontroller has CPU, RAM, ROM, peripherals (counter/timers, comparators, ADC, DC, DMA, etc.), and IO ports on one chip. Also, many new microcontrollers come with built-in oscillators too.

Chapter 2　Introduction to Embedded Robotics

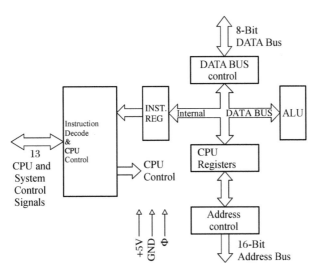

Figure 2-2　Z80 CPU Block Diagram

Essentially, a microcontroller gathers input, processes this information, and outputs a particular action based on the collected data. Microcontrollers usually operate at lower speed, around the 1MHz to 200 MHz range, and need to be designed to consume less power because they are embedded inside other devices that can have greater power consumption in other areas.

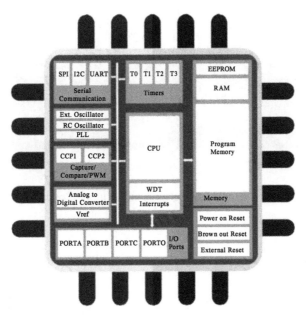

Figure 2-3　A Microcontroller

A typical microcontroller has a CPU, RAM, ROM, peripherals (counters/timers, comparators, ADC, DC, DMA, etc.), and IO ports on the same IC chip. Most modern MCU comes with internal oscillators, which means you can flash your code and hook up the DC power source, and voila! Everything is up and running!

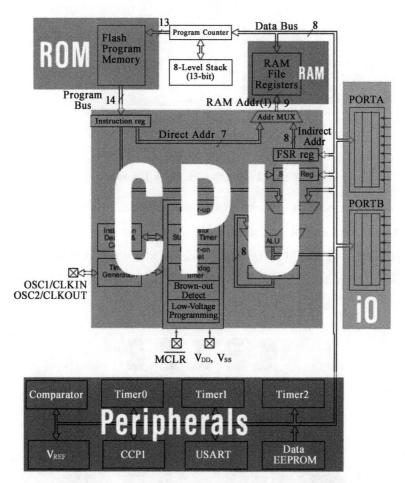

Figure 2-4　PIC16F628A Microcontroller Block Diagram

In addition to price, the most common factors that influence our choice of a microcontroller for a project are:

- Max CPU Frequency
- Memory Size (RAM)
- Bus Width (8-Bit, 16-Bit, 32-Bit)

Chapter 2 Introduction to Embedded Robotics

- IO Pins Count
- ADC (#Channels & Sampling Rate)
- DMA Channels
- Serial Ports (UART, SPI, I2C, USB, I2S, CAN)
- Other Peripherals (Timers, IRQ Pins, etc.)

Common MCUs include the Intel MCS-51, often referred to as an "8051 micro-controller," which was first developed in 1985; the AVR microcontroller designed by Atmel in 1996; the programmable interface controller (PIC) from Microchip Technology; and various licensed Advanced RISC Machines (ARM) microcontrollers.

Figure 2-5 Common MCUs

Another question is raised, which is

"Why don't we use computers, which are easier to programme for most novices and beginners, to Robotics Labs?"

To answer this question, we need to consider a few main points:

Microcontrollers are very efficient in terms of power consumption

- They are much cheaper.
- They have input/output ports.
- They are lighter and smaller.

A development board is a circuit board consisting of a CPU and minimal support circuits to develop an embedded system. Unlike a computer equipped with an advanced user interface to perform general tasks, a development board only has minimal hardware, emphasizing digital and analogue inputs and outputs to execute specific tasks. Microcontrollers are the central part of embedded systems and robotics.

Some Development Boards

While there are a variety of boards suitable for different projects, the following is a list of boards commonly used in robotics prototyping. This section aims to introduce some development boards for comparison, and it is impossible and beyond this book's scope to include all development boards. Furthermore, in this ever-expanding field, many new development boards may be added to the robotics market by the time this book is published. Similarly, discussing only one development board or the most popular one will deviate from the book's original goal of being a general handbook for Robotics Labs. In addition, an enormous number of resources are available online, and we must continuously call on those resources before we can complete any project.

Arduino Development Boards Family

Microcontrollers especially have very different current consumption in other states. What puts Arduino at the top of the microcontrollers list is

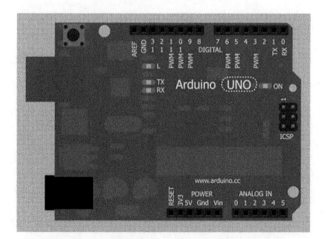

Figure 2-6 Arduino UNO Microcontroller Board

its open electronic licence, price, and a considerable number of online resources and community support. Arduino also is very low in power consumption. Arduino UNO development board using ATMega328 with the backing for sleep mode. ATMega328 in *Active Mode* will continuously execute several million instructions per second. However, forcing it into sleep mode makes it possible to reduce its current consumption to a dramatically low.

Arduino UNO

Arduino UNO is the most popular development board for both hobbyists and academics. It is number one regarding community support, available resources, and peripherals. The Arduino UNO is based on the ATmega328 microcontroller. It has 14 digital input/output pins (6 can be used as PWM outputs), six analogue inputs, a 16 MHz ceramic resonator, a USB connection, a power jack, an ICSP header, and a reset button[4]. It contains everything needed to support the microcontroller. The UNO differs from all preceding boards in that it does not use the FTDI USB-to-serial driver chip. Instead, it features the Atmega16U2 (Atmega8U2 up to version R2) programmed as a USB-to-serial converter. Today, Arduino UNO is not as powerful a development board as its competitors, but it still holds the number one choice for beginners to robotics.

Arduino family development boards and their programming IDE became standard for development boards within ten years of its birth in Italy. Its success is because it is open electronic, price, and simple structure and operation.

This board contains a USB interface, i.e. a USB cable to connect the board to a computer, and Arduino IDE (Integrated Development Environment) software to program the board.

The unit comes with 32 kB flash memory that stores the number of instructions, while the SRAM is 2 kB and EEPROM is 1 kB.

The associated circuitry operates at 5 V, while the input voltage ranges between 6 V to 20 V, and the recommended input voltage ranges from 7 V to 12 V.

As an open electronic, there are many clones of Arduino produced by different companies for various prices and quality. Some of those companies also offer a wide range of peripherals, shields as well as example software programs[5].

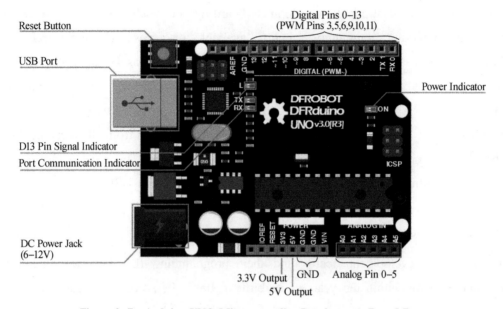

Figure 2-7　Arduino UNO Microcontroller Development Board Parts

Chapter 2　Introduction to Embedded Robotics

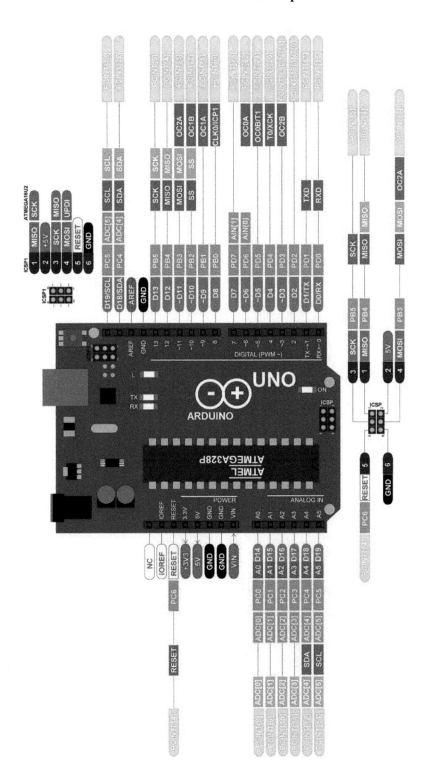

Figure 2 – 8　Arduino UNO Pinouts

Arduino Nano 3.3 BLE-sense

This is a new-born in the Arduino family, including integrated Bluetooth sensors, Accelerometer, Gyroscope, Manometer, Proximity, Light, Colour, Gesture, Temperature, Humidity, and Pressure sensors. Also, it has an onboard microphone[6].

This is a fully loaded Arduino which reduces use of extra devices and connections.

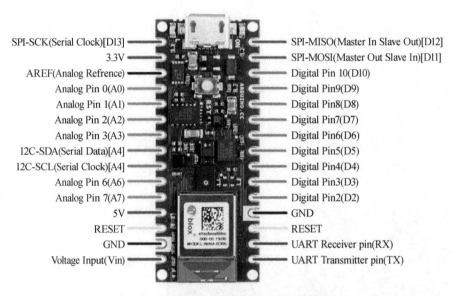

Figure 2-9　Arduino NANO 33 BLE-SENSE Pinouts

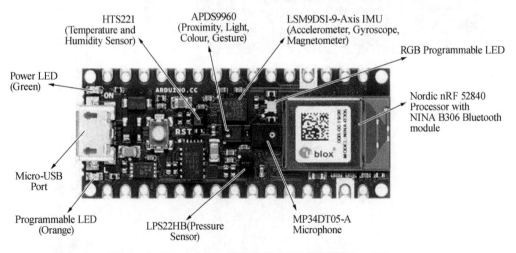

Figure 2-10　Arduino NANO 33 BLE-SENSE Board Spec

Chapter 2　Introduction to Embedded Robotics

Arduino Nano 33 has an onboard 9 axils

Accelerometer, Gyroscope, Manometer to provide enough information for a mobile robot where position data is crucial.

Related data are available out of box from example codes and ready to use with our projects. Example codes provide how to use on-board sensors and integrate them into our project codes.

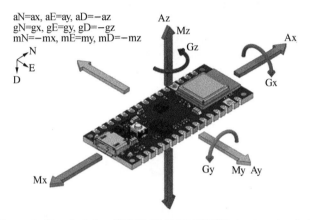

Figure 2-11　Arduino NANO 33 BLE-SENSE Acetometer Axis

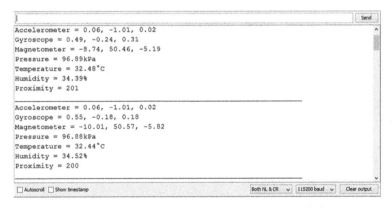

Figure 2-12　Arduino NANO 33 BLE-SENSE Acetometer's Output

Pocuter mini computer

The Pocuter is an Arduino® compatible mini-computer that combines a full colour OLED, a microcontroller with Wi-Fi and Bluetooth capability, Micro SD card connector for up to 512 Gigabytes of storage, three assignable buttons, microphone, accelerometer, temperature sensor, ambient light sensor with PPG capability, RGB LED and charging module in a coin sized package[7].

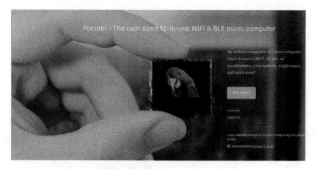

Figure 2-13　Pocuter

Hardware specification:

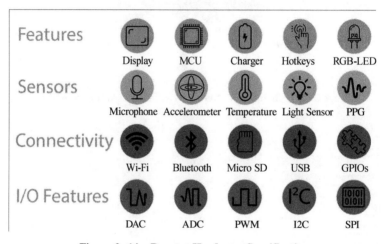

Figure 2-14　Pocuter Hardware Specification

Software overview:

Figure 2-15　Pocuter Software Specification

Chapter 2 Introduction to Embedded Robotics

The Pocuter is comparable to a 50 Cent coin. The ultra-compact size, together with all included features, enables excellent DIY projects!

The entire device is just 26 × 23 mm (roughly 1 × 0.9 inches). Together with the battery and housing, it would be about 11 mm (0.43 inches) thick.

The onboard full-colour OLED with 96 × 64 pixels inside a 0.95-inch panel can display bright and saturated 16-Bit colours!

256 KB of internal Flash and 32 KB SRAM with an ESP32-CS, enough processing power to utilize everything on the board—all at once.

The ESP32 adds Wi-Fi & Bluetooth to the Pocuter. This makes it much more powerful, together with the Pocuter API & Cloud! (Read the "Software & SDK" section to learn more about it!)

An onboard electret condenser microphone, also widely used in smartphones, enables voice commands, dictating text and recording voice. Together with a speaker, it would even allow calls.

The 3-axis accelerometer enables the Pocuter to detect motion and which way it is tilted. A shake and specific tilt detection enables automatic wakeups.

The temperature sensor is built into the same IC as the accelerometer, enabling you to measure the current temperature linearly.

STM32

The STM32 board, Blue Pill, is a development board for the ARM Cortex M3 Microcontroller. It looks very similar to the Arduino Nano but packs in quite a punch.

These boards are incredibly cheap compared to the official Arduino boards, and the hardware is open-source. The microcontroller on top of it is the STM32F103C8T6 from STMicroelectronics. Apart from the microcontroller, the board also holds two crystal oscillators: an 8 MHz crystal and a 32 KHz crystal, which can be used to drive the internal RTC (Real Time Clock). Because of this, the MCU can operate in deep sleep modes, making it ideal for battery-operated applications[8].

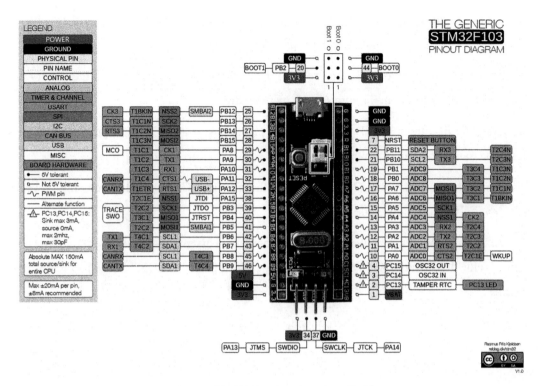

Figure 2-16　STM32 Development Board's Pinouts

Some of STM32 features:

- *Architecture*: 32-bit ARM Cortex M3
- *Operating Voltage*: 2.7 V to 3.6 V
- *CPU Frequency*: 72 MHz
- *Number of GPIO pins*: 37
- *Number of PWM pins*: 12
- *Analog Input Pins*: 10 (12-bit)
- *USART Peripherals*: 3
- *I²C Peripherals*: 2
- *SPI Peripherals*: 2
- *Can 2.0 Peripheral*: 1
- *Timers*: 3(16-bit), 1 (PWM)
- *Flash Memory*: 64 KB
- *RAM*: 20 KB

Chapter 2　Introduction to Embedded Robotics

Raspberry PI Pico

Raspberry Pi Pico with 32-bit ARM Cortex M0+ and 264 Kbyte SRAM memory is a choice both for beginners and professionals[9]. Raspberry Pi Pico is a low-cost, high-performance microcontroller board with flexible digital interfaces. Programmable in C and Micro Python, Pico is adaptable to a vast range of applications and skill levels, and getting started is as easy as dragging and dropping a file.

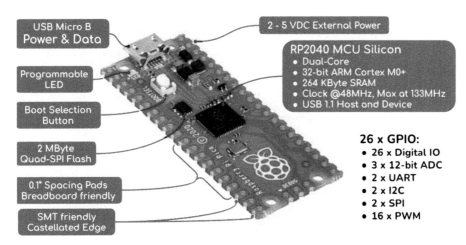

Figure 2-17　Raspberry Pi Pico Details

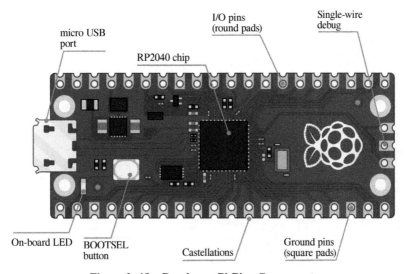

Figure 2-18　Raspberry Pi Pico Components

Raspberry Pi Pico Specification:

Raspberry Pi Pico as a new development board has a rich set of features at a very low price.

- Dual ARM Cortex-M0+ @ 133 MHz
- 264 KB on-chip SRAM in six independent banks
- Support for up to 16 MB of off-chip flash memory via dedicated QSPI bus
- DMA controller
- Fully-connected AHB crossbar
- Interpolator and integer divider peripherals
- On-chip programmable LDO to generate core voltage
- 2 on-chip PLLs to generate USB and core clocks
- 30 GPIO pins, 4 of which can be used as analogue inputs
- Peripherals
- 2 UARTs
- 2 SPI controllers
- 2 I2C controllers
- 16 PWM channels
- USB 1.1 controller and PHY, with host and device support
- 8 PIO state machines

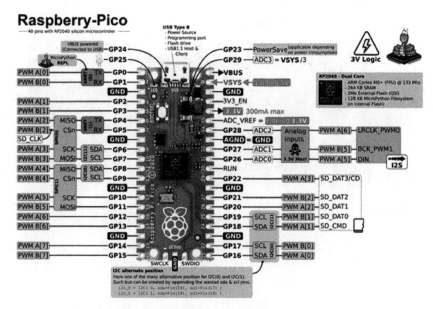

Figure 2-19　Raspberry Pi Pico-pinouts

Chapter 2   Introduction to Embedded Robotics

Figure 2-20   I2C Communication with Raspberry Pi Pico

Teensy Board

This 32-bit microcontroller board makes it easy to get started with the Teensy family of products.

With the Teensy LC's built-in USB port and flashed bootloader, you can easily upload new code. There is no need for a third-party programmer. You can use your favourite C-based IDE to write programmes for the Teensy, or you can download the Teensyduino extension for the Arduino IDE to create "sketches for Teensy!"[10]

The Teensy's processor can act as an emulated version of any USB device, making it ideal for USB-MIDI and other HID applications.

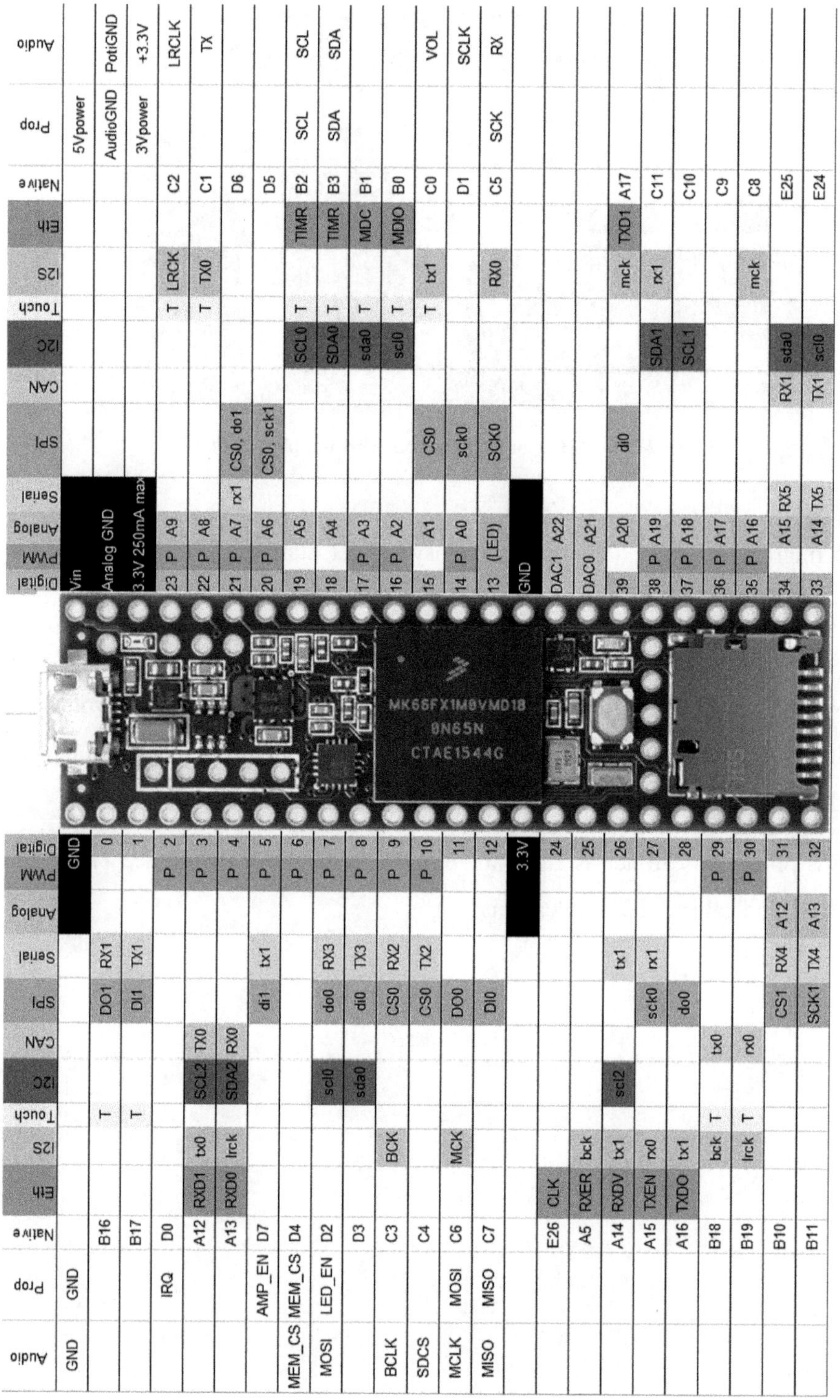

Figure 2−21 Teensy Board-pinouts

Pyboard

The pyboard is the official MicroPython[11] microcontroller board with full support for software features. The hardware has:

MicroPython uses C99 for its code and makes its entire core available to the public under the highly permissive MIT licence.

Some third-party libraries and extension modules are included, but most are open source and licenced under MIT or a similar licence.

MicroPython is open-source software, meaning you are free to use it in your own projects, classrooms, or even for profit.

Various HID-related and MIDI-related initiatives. MicroPython is developed in the open on GitHub and the source code is available at the GitHub page, and on the download page.

Figure 2-22 Pyboard

Bluno Beetle and Bluno Remote

If you're a software or hardware developer interested in wireless prototyping, the Bluno board can be your first choice. Bluno is the first Arduino board to integrate a BT 4.0 (BLE) module[5]. Create your own wearable electronic devices like a smart bracelet, pedometer, and more!

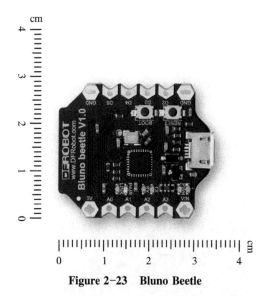

Figure 2-23 Bluno Beetle 　　　　Figure 2-24 Bluno Remote

Also, Bluno Remote is equipped with Bluetooth for wireless connectivity.

Pinout Diagram

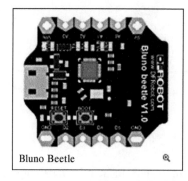
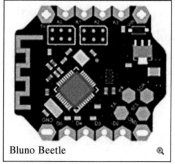

Figure 2-25 Bluno Beetle Pinout

- Pin Mapping

Table 2-1 Bluno Beetle Pinout

Sikscreen	Digtal Pin	PWM Channel	Analog Channel	UART	I2C
RX	0			Serial 1	
TX	1				
SOA	A4				SOA
SCL	A5				SCL

(Continued)

Sikscreen	Digtal Pin	PWM Channel	Analog Channel	UART	I2C
D2	2				
D3	3	3			
D4	4				
D5	5	5			
A0	A0		A0		
A1	A1		A1		
A2	A2		A2		
A3	A3		A3		

- Power Interface Description

Table 2-2 Bluno Beetle Power Interface

Sikscreen	Description
VIN	external power supply < 8 V
5 V	5 V positive supply
GND	GND

Example Circuit Using Bluno Beetle

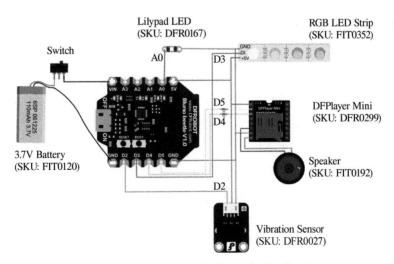

Figure 2-26 Bluno Beetle Example Application

Bluno Beetle proves that robotics control does not depend on the controller size. It is possible to run DC motors using Blunu Beetle too, and it is an excellent choice for tiny-size prototyping.

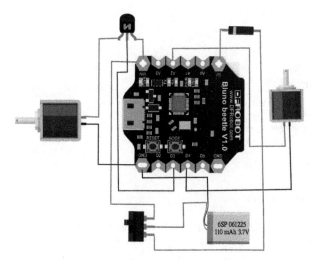

Figure 2-27　Bluno Beetle Using DC Motor with PWM

ESP32 + OLE + Wi-Fi

ESP32 with OLED and Wi-Fi, also is considered as hi performance development when small size and low power are our key importance[12].

- Operating voltage: 3.3 V to 7 V
- Operating temperature range: -40 ℃ to 90 ℃
- Data rate: 150 Mbps @ 11n HT40, 72 Mbps @ 11n HT20, 54 Mbps @ 11g, 11 Mbps @ 11b
- Transmit power: 19.5 dBm @ 11b, 16.5 dBm @ 11g, 15.5 dBm @ 11n
- Flash size: 4MByte (32M-bits)
- Module size: 50 ×25.4 ×10.3 mm
- Lithium battery interface: 2Pin-1.25mm interface
- Receiver sensitivity up to -98 dBm
- UDP continues to increase throughput by 135 Mbps
- Supports Sniffer, Station, softAP and Wi-Fi Direct modes

Chapter 2   Introduction to Embedded Robotics

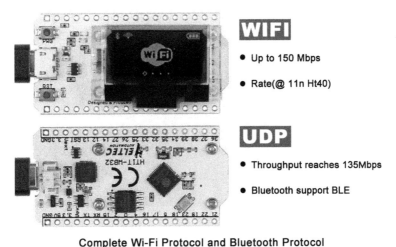

Figure 2-28   ESP32 + OLED + Wi-Fi

Resource	Parameter	
Master Chip	ESP32(240MHz Tensilica LX6 dual-core + 1 ULP, 600 DMIPS)	
Wireless Communication	Wi-Fi	Bluetooth
	802.11 b/g/n (802.11n up to 150 Mbps)	Bluetooth V4.2 BR/EDR and Bluetooth LE specification
Hardware Resource	UART x 3; SPI x 2; I2C x 2; I2S x 1; 12-bits ADC input x 18; 8-bits DAC output x 2; GPIO x 22, GPI x 6	
FLASH	4MB(64M-bits) SPI FLASH	
RAM	520KB internal SRAM	
Interface	Micro USB x 1; 18 x 2.54 pin x 2	
Maximum Size (Including protruding parts such as switch and battery compartment)	51 x 25.5 x 10.6 mm	
USB to Serial Chip	CP2102	
Battery	3.7V Lithium(SH1.25 x 2 socket)	
Solar Energy	x	
Battery Detection Circuit	√	
External Device Power Control (Vext)	√	
Display Size	0.96-inch OLED	
Working Temperature	-40~80℃	

Figure 2-29   ESP32 Specification

OpenCR

OpenCR is a powerful embedded development board suitable for many robot bases, especially if Robotis servo motors are in use.

OpenCR1.0 is developed for ROS-embedded systems to provide utterly open-source hardware and software. The STM32F7 series chip inside the OpenCR1.0 board is based on a powerful ARM Cortex-M7 with a floating point unit[13].

The development environment for OpenCR1.0 is wide open, from Arduino IDE and Scratch for young students to traditional firmware development for the experts.

Figure 2-30 OpenCR Development Board

OpenCR Specification:

Table 2-3 OpenCR Board Specification

Items	Specifications
Microcontroller	STM32F746ZGT6/32-bit ARM Cortex®-M7 with FPU (216MHz, 462DMIPS)
Sensors	(Discontinued) Gyroscope 3Axis, Accelerometer 3Axis, Magnetometer 3Axis (MPU9250) (New) 3-axis Gyroscope, Axis Accelerometer, A Digital Motion Processor™ (ICM-20648)
Programmer	ARM Cortex 10pin JTAG/SWD connector USB Device Firmware Upgrade (DFU) Serial
Digital I/O	32 pins (L 14, R 18) * Arduino connectivity 5Pin OLLO ×4 GPIO ×18 pins PWM ×6 I2C ×1 SPI ×1
Analog INPUT	ADC Channels (Max 12 bit) ×6
Communication Ports	USB ×1 (Micro-B USB connector/USB 2.0/Host/Peripheral/OTG) TTL ×3 (B3B-EH-A/DYNAMIXEL) RS485 ×3 (B4B-EH-A/DYNAMIXEL) UART ×2 (20010WS-04) CAN ×1 (20010WS-04)

Chapter 2　Introduction to Embedded Robotics

(Continued)

Items	Specifications
LEDs and Buttons	LD2 (red/green): USB communication User LED × 4: LD3 (red), LD4 (green), LD5 (blue) User button × 2-Power LED: LD1 (red, 3.3 V power on) Reset button × 1 (for power reset of board) Power on/off switch × 1
Input Power Sources	5 V (USB VBUS), 5–24 V (Battery or SMPS) Default battery: LI-PO 11.1V 1,800mAh 19.98Wh Default SMPS: 12V 4.5A External battery Port for RTC (Real Time Clock) (Molex 53047-0210)
Input Power Fuse	125 V 10A Little Fuse 0453010
Output Power Sources	* 12 V max 4.5 A(SMW250-02) * 5 V max 4 A(5267-02A), 3.3 V@ 800 mA(20010WS-02)
Dimensions	105(W) × 75(D) mm
Weight	60 g

Note: * 5 V power source is supplied from regulated 12 V output. Total power consumption on 12 V and 5 V ports should not exceed 55 W.

OpenCR includes a connector that is compatible with the Arduino Uno pin map.

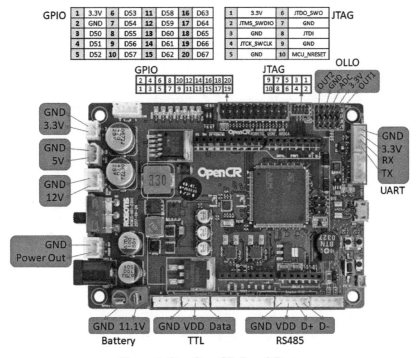

Figure 2-31　OpenCR Board Parts

The pins 0 to 21 are the same pin as the Arduino Uno; after that, they are mapped to the pins added toOpenCR.

Figure 2-32 OpenCR Board-Arduino UNO Support

To prepare the OpenCR board for use with Arduino IDE, we need to make its USB connection to be visible to Arduino and can compile Arduino codes. For this:

(in Linux terminal)

$ wget https://raw.githubusercontent.com/ROBOTIS-GIT/OpenCR/master/99-opencr-cdc.rules

$ sudo cp ./99-opencr-cdc.rules/etc/udev/rules.d/

$ sudo udevadm control-reload-rules

$ sudo udevadm trigger

And since the OpenCR libraries are built for 32-bit platform, 64 bit PC needs the 32 bit compiler relevant for the Arduino IDE.

$ sudo apt-get install libncurses5-dev:i386

More information about OpenCV board from:

https://emanual.robotis.com/docs/en/parts/controller/opencr10/

OpenCM

Another development board from Dynamixel is OpenCM, a much smaller version of OpenCR yet very powerful, mainly to control some Dynamixel Servo Motors. OpenCM9.04 is a microcontroller board based on 32-bit ARM Cortex-M3. The OpenCM9.04's schematics and source codes are open-source. OpenCM is also compatible with Arduino IDE. There are three types of OpenCM:

Item	Description
CPU	STM32F103CB (ARM Cortex-M3)
Operation Voltage	5 – 16V
I/O	GPIO x 26
Timer	4 (16bit)
Analog Input(ADC)	10 (12bit)
Flash	128KB
SRAM	20KB
Clock	72Mhz
USB	1 (2.0 Full Speed) Micro B Type
USART	3
SPI	2
I2C(TWI)	2
Debug	JTAG & SWD
DYNAMIXEL TTL BUS	4 (Max 1Mbps)
Dimensions	27mm x 66.5mm

Figure 2-33 OpenCM Specification

Note:

USB power cannot operate Dynamixel servo motors attached to the OpenCM board. A separate power supply must be used.

[OpenCM9.04 A-Type]　　　　[OpenCM9.04 B-Type]　　　　[OpenCM9.04 C-Type]

Figure 2-34　OpenCM Boards

(OpenCM9.04 can operate using power supplied via USB, battery, + - terminal.)

Check the operating voltage for peripheral devices when using a power supply. Dynamixel XL-series receives the same voltage.

XL-320 cannot be used with other Dynamixel Servo motors due to the difference in operating voltages.

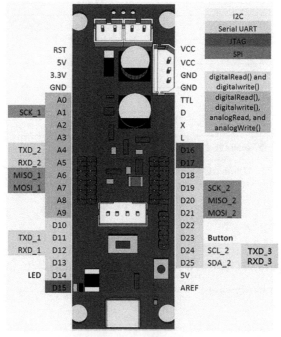

Figure 2-35　OpenCM-pinouts

Chapter 2 Introduction to Embedded Robotics

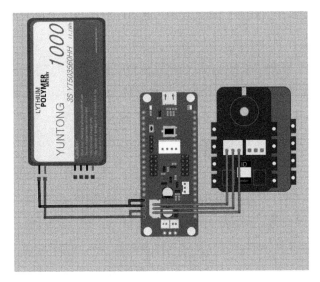

Figure 2-36 OpenCM-servo and Power Connections

Similar to OpenCR, there are some setting and installation processes to use OpenCM with Arduino IDE: (in Linux terminal)

$ Wget https://raw. githubusercontent. com/ROBOTIS-GIT/OpenCM9. 04/master/99-opencm-cdc. rules

$ sudo cp. /99-opencm-cdc. rules/etc/udev/rules. d/

$ sudo udevadm control—reload-rules

$ sudo udevadm trigger

And to support using 64 bin PCs:

$ sudo apt-get install libncurses5-dev: i386

More information regarding OpenCM boards:

https://emanual. robotis. com/docs/en/parts/controller/opencm904/#opencm-ide

Development Boards' Features

Flash Memory

Unlike random access memory (RAM), data stored in flash memory can be kept for a long time, even when the microcontroller is powered down.

Forever storing the microcontroller's uploaded programme, this is a must-have.

EPROM, EEPROM

EEPROM is similar to Flash Memory in that it is non-volatile and keeps its data even when the power is turned off. EEPROM differs from Flash Memory in that it allows for the rewriting of individual bytes rather than entire "blocks". This will expand the life of EEPROM in contrast to Flash Memory.

Serial Bus

Serial Bus Interface is the serial communication in the microcontroller, which sends data one bit at a time. Microcontroller boards connect ICs with signal traces on a printed circuit board (PCB). Using a serial bus to transfer data can reduce the number of pins in ICs, making them cheaper. Examples of serial buses in ICs are SPIs or I2Cs.

Serial Peripheral Interface (SPI)

Motorola's microcontrollers typically feature the Serial Peripheral Interface (SPI), a synchronous serial bus that Motorola designed and implemented. The signals that make up the SPI bus are the active-low slave select (SS), serial clock (SCK), master out (MOSI), and master in (MISO) signals.

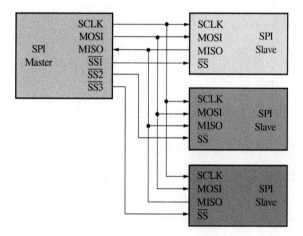

Figure 2-37　SPI Connections

Chapter 2 Introduction to Embedded Robotics

I2C Serial Bus

The goal of the Inter-Integrated Circuit (I2C) Protocol is to facilitate communication between several "peripheral" digital integrated circuits ("chips") and a controller chip or chips. I2C, like its predecessor Serial Peripheral Interface (SPI), is designed for local communications between components of a single electronic system [14].

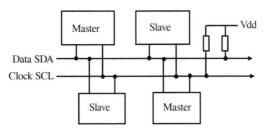

Figure 2-38 I2C Connections

In robotics projects, I2C is commonly used where I/Os are insufficient to connect many devices. However, there are other reasons why we may use I2C. In general, we may choose I2C communication over asynchronous because:

1—Devices using them must agree on a data rate ahead of time, and the two devices must also have clocks close to the same rate.

2—Asynchronous serial ports require hardware overhead—the UART at either end is relatively complex and challenging to implement in software if necessary accurately. I2C needs one start and stop bit is a part of each frame of data, which means 10 bits of transmission time are required for every 8 bits of data sent.

3—Asynchronous serial ports are suitable for communications between two and only two devices. Connecting more than two devices to the serial bus may be possible, but that is not always free of issues and problems.

4—Theoretically, there is no limit to asynchronous serial communications, but most UART devices only support a particular set of fixed baud rates, with the highest usually around 230400 b/s. In robotics projects, I2C is commonly used.

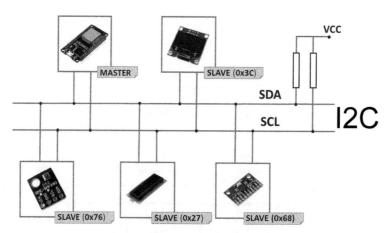

Figure 2-39 I2C Multiple Device Connections

CAN Bus

To facilitate application-to-application communication between microcontrollers and devices in vehicles, the Controller Area Network (CAN Bus) standard was developed where there is no host. It's a message-based protocol initially developed to help reduce copper usage in automotive electrical wiring, but it has found many other applications beyond that.

The data in a frame is sent serially from one device to another, but in such a way that the device with the highest priority can keep sending while the others stop. Every device, including the one doing the transmitting, receives frames.

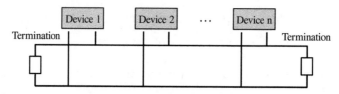

Figure 2-40 CAN Bus Connections

PWM (Pulse Width Modulation)

A pulse-width-modulated (PWM) signal is a digital square wave in which the frequency remains fixed but the amount of time the signal is "on" (the duty cycle) can change from 0% to 100%. PWM is used to generate analogue output and, therefore, to control devices where an analogue signal is needed. PWM is one of the essential features of development boards and will be discussed in detail in section III of this chapter.

Digital I/O

The input pins on a microcontroller allow it to take in data from the outside world, and digital or analogue signals may be input via the pins. In contrast to digital outputs, digital inputs require an external source to set their voltage to either 0 or 5 (or 3.3) volts. A bit in the microcontroller's memory stores the value of each digital input.

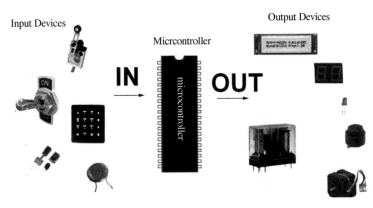

Figure 2-41 Input/Output Devices

Also, digital output ports are provided with microcontroller boards to connect to output devices for control, visualisation of output, or just as a primary user interface. Software digital outputs can be set to either 0 volts as a LOW signal or 5 (or 3.3) volts as a HIGH signal.

Analog I/O

In contrast to a digital signal, an analogue signal can take any number and is not limited to 'LOW' and 'HIGH'. Some development boards like Arduino have a built-in analogue-to-digital converter (ADC) to measure the value of analogue signals. As discussed in Chapter one, the ADC turns the analogue voltage into a digital value. Arduino uses analogRead(pin x) to use a reference voltage as a basis for converting the input voltage to a digital value between 0 and 1023. 5 V or 3.3 V is the reference voltage by default (for 3.3 V Arduino boards). Arduino Uno, therefore, has an analogue read resolution of 1024. This can be varied in other development boards.

Similarly, a digital analogue converter (DAC), is used to produce an output analogue signal. Some boards don't have DAC, such as Arduino, but they can use Pulse Width

Modulating (PWM), a digital signal to achieve some of the functions of an analogue output. Arduino uses 'analogWrite (pin, value)' to produce an analogue signal. Where the pin is a PWM enable pin, the value is between 0 to 255, representing voltage levels from 0 volts to 5 (3.3) volts[1].

An example of using the 'analogWrite' function in Arduino is controlling the brightness of a LED[1]. HIGH signal in some microcontrollers is 3.3 volt[15].

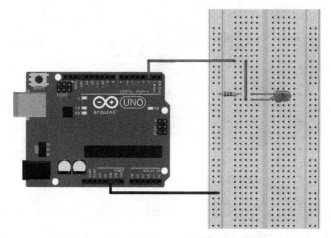

Figure 2-42   Controlling Brightness of LED

```
01.  const int pwm = 2 ;         //initializing pin 2 as 'pwm' variable
02.  void setup()
03.  {
04.      pinMode(pwm,OUTPUT) ;   //Set pin 2 as output
05.  }
06.  void loop()
07.  {
08.      analogWrite(pwm,25) ;       //setting pwm to 25
09.      delay(50) ;                 //delay of 50 ms
10.      analogWrite(pwm,50) ;
11.      delay(50) ;
12.      analogWrite(pwm,75) ;
13.      delay(50) ;
14.      analogWrite(pwm,100) ;
15.      delay(50) ;
16.      analogWrite(pwm,125) ;
17.      delay(50) ;
18.      analogWrite(pwm,150) ;
19.      delay(50) ;
20.      analogWrite(pwm,175) ;
21.      delay(50) ;
22.      analogWrite(pwm,200) ;
23.      delay(50) ;
24.      analogWrite(pwm,225) ;
25.      delay(50) ;
26.      analogWrite(pwm,250) ;
27.  }
```

Interrupts

An interrupt signal pushes the processor to immediately stop what it is doing and handle some high-priority processing. The high-priority processing is called an Interrupt Handler.

An interrupt handler is like any other void function. If you write one and attach it to an interrupt, it will get called whenever that interrupt signal is triggered. When the process returns from the interrupt handler, it goes back to continue what it was doing before.

There are many reasons we want to use an interrupt. For example, an obstacle detection that must be detected in more priorities or a wheel encoder to check the speed of wheels. When an interrupt occurs, the main programme stops running so that another routine can be executed.

After this subroutine is complete, control is handed back to the main programme.

Type of Interrupts:

- Hardware Interrupts—This kind of interrupt is a response to an external event (HIGH or LOW).
- Software Interrupts—When an instruction is sent in the software.

How Arduino board handling interrupts:

Arduino language only supports the "attachInterrupt()" function using the following three constants:

- LOW to trigger the interrupt to respond to a LOW signal on the pin.
- CHANGE to trigger the interrupt to respond to a change in the state of the pin.
- FALLING to trigger the interrupt when the pin goes from high to low.

Interrupts can be generated from several sources:

- The timer interrupts from development board's timers.
- External Interrupts from a change in the state of one of the external interrupt pins.
- Pin-change interrupts a change in the state of any one of a group of pins.

```
1.  int pin = 2; //define interrupt pin to 2
2.  volatile int state = LOW; //This is the initial state before calling
3.                                            // interrupt function
4.
5.  void setup() {
6.      pinMode(13, OUTPUT);
7.      attachInterrupt(digitalPinToInterrupt(pin), blink, CHANGE);
8.      //calls interrupt function blink() when pin 2 state state is changed
9.  }
10. void loop() {
11.     digitalWrite(13, state); //pin 13 equal the state value
12. }
13.
14. void blink() {
15.     //This is interrupt function
16.     state = !state; //toggle the state when the interrupt occurs
17. }
```

Figure 2-43 An Example Code for Interrupts Using Arduino Uno Development Board

Each development board may have a different number of interrupts. For instance:

OpenCV board has external interrupts Ø to 15 on pins (DØ, D1, D2, D2, D3, D4, D5, D6, D7, D8, D9, D10, D11, D12, D13, D14, D15)

With each microcontroller, specific pins can be used for external interrupts and can be used with attachInterrupt (EXTI_Pin, callbackFunction, Mode) function.

Most Arduino boards only have hardware interrupts (interrupt 0 and interrupt 1-pin 2 and 3). Arduino Mega has six hardware interrupts on pins 21, 20, 19 and 18. To find available interrupts on different development boards, we need to refer to the board's pinouts.

Development Boards and Power Consumption

Microcontroller boards within the same family may have completely different current consumption. Because it draws about 45 mA from its 9 V power supply, an Arduino Uno has a lifespan of less than a day. If you use an Arduino Pro Mini to run on either 3.3 or 5 volts, its power consumption drops to 54μA (0.054 mA) in sleep mode and 23 μA (0.023 mA) in power-down mode. That is about four years on a 9 V battery with 1,200 mAh capacity, or 2,000 times more efficient than the Arduino Uno. By removing the voltage regulator, the power consumption is only 4.5 μA for the 3.3 V version and 5.8 μA for the 5 V version in power-down sleep[2].

Chapter 2　Introduction to Embedded Robotics

Figure 2-44　Development Boards and Power Consumption

Table 2-4　Arduino Uno Technical Specifications

Microcontroller	ATmega328P—8 bit AVR family microcontroller
Operating Voltage	5 V
Recommended Input Voltage	7–12 V
Input Voltage Limits	6–20 V
Analog Input Pins	6(A0–A5)
Digital I/O Pins	14 (out of which 6 provide PWM output)
DC Current on I/O Pins	40 mA
DC Current on 3.3 V Pin	50 mA
Flash Memory	32 KB (0.5 KB is used for bootloader)
SRAM	2 KB
EEPROM	1 KB
Frequency (Clock Speed)	16 MHz

Source: https://components101.com/microcontrollers/arduino-uno

Providing an Arduino Uno development board with the required power cannot guarantee the proper operation of peripherals and attached devices if Arduino powers them. The maximum current allowed from a GPIO on an ATmega328 microcontroller is 40 mA. In addition, the total current through the supply or ground rails (i.e. the sum of all current OP wants the GPIO pins to sink or source) is limited to 200 mA.

```
Operating Temperature..........................−55°C to +125°C
Storage Temperature............................−65°C to +150°C
Voltage on any Pin except RESET
with respect to Ground ........................−0.5V to $V_{CC}$+0.5V
Voltage on RESET with respect to Ground......−0.5V to +13.0V
Maximum Operating Voltage ............................. 6.0V
DC Current per I/O Pin ................................. 40.0 mA
DC Current $V_{CC}$ and GND Pins............................. 200.0 mA
```

Figure 2−45   Some of ATmega328P Specification
Source: http://www.atmel.com/Images/doc8161.pdf

If we use a 5 V pin to power Arduino which is not going through the microcontroller, then Arduino can provide larger power to support peripherals. The USB interface places a cap of 500 mA on the total current that can be sent to an Arduino when powering over USB. If you power your Arduino from an external source, the maximum power provided through a 5 V pin is 1 A. However, this is also *thermally limited*, meaning that as you draw power, the regulator will heat up. When it overheats, it will shut down temporarily[3].

SBC (Single Board Computer)

Raspberry Pi

The Raspberry Pi is a small single-board computer (SBC) created by the Raspberry Pi Foundation and Broadcom in the United Kingdom.

The original focus of the Raspberry Pi project was to encourage the teaching of computer science fundamentals in K-12 institutions and low-income countries. However, from the first

Chapter 2 Introduction to Embedded Robotics

Figure 2-46　Raspberry PI Board

series, it became very popular, especially in schools. Since then, raspberry pi has gained more reputation.

Raspberry Pi, in general, is one of the most reliable, affordable single-board computers ever made. If Arduino became a standard in development boards, raspberry pi would be a standard in single board computers in education and small robotics projects. However, Raspberry Pi CM4 has recently opened its way to the industry.

Although many operating systems (Linux) are available for Raspbian Pi (even Ubuntu), Raspbian, the original Linux Operating System developed by Raspberry Pi, is stable and reliable enough for most applications.

There are three distinct versions of Raspberry Pi, and each has seen multiple revisions. Broadcom's system-on-a-chip (SoC) integrates an ARM-compatible CPU and graphics processing unit (GPU) for use in Raspberry Pi single-board computers (SBCs), while the RP2040 SoC in Raspberry Pi Pico accomplishes ARM-compatible central processing unit (CPU).

Figure 2-47　Raspberry Pi Zero

There is a massive bank of resources available for raspberry pi for every field of robotics.

Perhaps Raspberry Pi zero is one of the most miniature single board computers that, without any hassle, can run Python, OpenCV, and ROS (Robot Operating System).

There is a built-in reference can be accessed from a terminal on the Raspberry Pi and running the command pinout. This tool is available on the Raspberry Pi OS desktop image, but not on Raspberry Pi OS Lite.

More resources are available from:

https://www.raspberrypi.com/documentation/computers/raspberry-pi.html

Chapter 2　Introduction to Embedded Robotics

PROJECT: Linux Box

Linux Box v1.0

Linux Box is a full computer based on Raspberry Pi, especially built in our robotics lab to make developing robotics projects easier. Linux Box is a project under development to make it a perfect choice for small robotics and IoT projects

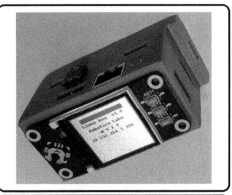

Specification:
—1 GHz Quad Core, 64-bit Arm Cortex CPU
—512 MB SDRAM
—2.4 GHz Wi-Fi
—Bluetooth 4
—128 GB SSD storage
—SD card
—3 full size USB2
—1 Mini HDMI
—4 hours of battery power
—Charging connection
—On/Off switch
—Online OLED 1.4" screen
—Joystick
—3 × programmable switch

Software:
—Raspbian based Linux
—ROS Kinetic
—OpenCV
—Python 2.7, 3.6
—Battery monitoring
—OLED support
—Arduino packages
—Many robotics tools

Figure 2-48　Linux Box

GPU Powered Boards

One of the early popular AI-powered development boards was Jetson boards. Jetson TK1, then Jetson TX1, and with move memory version Jetson TX2.

However, Jetson Nano is a smaller version that became a Raspberry Pi competitor with GPU and AI ready.

The Jetson Xavier series have even gone higher in performance than TX2, yet still in small sizes.

Nvidia Jetson TX2

Figure 2-49　Jetson TX2

The NVIDIA Jetson TX2 NX is the next evolution in AI performance for affordable embedded and edge products.

It's perfect for your next AI solution in any industry, from manufacturing and retail to agriculture and the life sciences, thanks to its small size and low power consumption[16].

You can quickly bring your best product to market with the assistance of pre-trained AI models, the Transfer Learning Toolkit, and the NVIDIA JetPack Software Development Kit.

Table 2-5　Jetson TX2 Specifications

AI performance	1.33 TFLOPS
GPU	NVIDIA Pascal architecture with 256 NVIDIA CUDA cores

(Continued)

CPU	Dual-core NVIDIA Denver 2 64-bit CPU and quad-core Arm Cortex-A57 MPCore processor complex
Memory	4 GB 128-bit LPDDR4 51.2 GB/s
Storage	16 GB eMMC 5.1
Power	7.5 W ∣ 15 W
PCIe	1 ×1 + 1 ×2 PCIe Gen2, total 30 GT/s
CSI Camera	Up to 5 cameras (12 via virtual channels) 12 lanes MIPI CSI-2 (3 ×4 or 5 ×2) D-PHY 1.2 (up to 30 Gbps)
Video encode	1 ×4K60 ∣ 3 ×4K30 ∣ 4 × 1080p60 ∣ 8 × 1080p30 (H.265) 1 ×4K60 ∣ 3 ×4K30 ∣ 7 × 1080p60 ∣ 14 × 1080p30 (H.264)
Video decode	2 ×4K60 ∣ 4 ×4K30 ∣ 7 × 1080p60 ∣ 14 × 1080p30 (H.265 & H.264)
Display	2multi-mode DP 1.2/eDP 1.4/HDMI 2.0 1x 2 DSI (1.5Gbps/lane)
Networking	10/100/1000 BASE-T Ethernet
Mechanical	69.6 mm ×45 mm 260-pin SO-DIMM edge connector

Nvidia Nano run modern AI workloads.

AI frameworks and models for tasks like image classification, object detection, segmentation, and speech processing are now accessible to developers, learners, and makers.

The NanoTM Developer Kit (V3) is a small, power-efficient (using as little as 5 Watts) and inexpensive platform that can handle modern AI workloads.

Figure 2-50　Jetson Nano

Developers, students, and makers can use artificial intelligence frameworks and models for tasks such as image classification, object detection, segmentation, and speech processing.

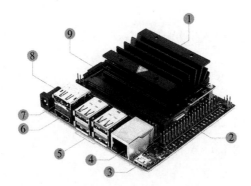

1. Micro SD card slot: insert a 16GB or larger TF card for main storage and writing system image;
2. 40-pin expansion header;
3. Micro USB port: for 5 V power input or for USB data transmission;
4. Gigabit Ethernet port: 10/100/1000 Base-T auto-negotiation;
5. 4 × USB 3.0 port;
6. HDMI output port;
7. DisplayPort connector;
8. DC jack: for 5 V power input;
9. MIPI CSI camera connector.

Figure 2-51　Jetson NANO Board Parts

Jetson Nano Developer Kit Features

- GPU: 128-core Maxwell™ GPU
- CPU: quad-core ARM® Cortex®-A57 CPU
- Memory: 4GB 64-bit LPDDR4
- Storage: Micro SD card slot (requires an external minimum 16G TF card)
- Video:
  - Encode: 4K @ 30 (H.264/H.265)
  - Decode: 4K @ 60 (H.264/H.265)
- Interfaces:
  - Ethernet: 10/100/1000BASE-T auto-negotiation
  - Camera: 12-ch (3 ×4 OR 4 ×2) MIPI CSI-2 DPHY 1.1 (1.5Gbps)
  - Display: HDMI 2.0, DP (DisplayPort)
  - USB: 4 × USB 3.0, USB 2.0 (Micro USB)
  - Others: GPIO, I2C, I2S, SPI, UART

Chapter 2　Introduction to Embedded Robotics

- Power:
  - Micro USB (5 V 2 A)
  - DC jack (5 V 4 A)
- Dimensions:
  - Core module: 69.6 mm × 45 mm
  - Whole kit: 100 mm × 80 mm × 29 mm

Table 2-6　Jetson Nano Specifications

	Jetson Nano	Jetson TX2
AI Performance	472 GFLOPs	1.3 TFLOPs
GPU	128-core NVIDIA Maxwell™ GPU	256-core NVIDIA Pascal™ GPU
CPU	Quad-Core ARM® Cortex®-A57MPCore	Dual-Core NVIDIA Denver 2 64-Bit CPU and Quad-Core ARM® Cortex®-A57MPCore
Memory	4 GB 64-bit LPDDR4 Memory	8 GB 128-bit LPDDR4 Memory
Storage	Micro SD card slot	32 GB eMMC 5.1
Video	Encode: 4K @ 30 (H.264/H.265) Decode: 4K @ 60 (H.264/H.265)	Encode: 4K × 2K 60 Hz (HEVC) Decode: 4K × 2K 60 Hz (12-bit support)
Connectivity	Gigabit Ethernet	Gigabit Ethernet, Wi-Fi, Bluetooth
CSI	12 × CSI-2 D-PHY 1.1 (Up to 18 GB/s)	12 × CSI-2 D-PHY 1.2 (Up to 30 GB/s)
Display	Two Multi-Mode DP 1.2 eDP 1.4 HDMI 2.0 1 × 2 DSI (1.5 Gbps/lane)	Two Multi-Mode DP 1.2 eDP 1.4 HDMI 2.0 Two 1 × 4 DSI (1.5 Gbps/lane)
PCIE	Gen 2	1 × 4	Gen 2	1 × 4 + 1 × 1 OR 2 × 1 + 1 × 2
Power	5 W/10 W	7.5 W/15 W

Jetson Xavier NX

The NVIDIA Jetson Xavier NX is a compact system-on-module (SOM) that delivers supercomputer performance at the edge. Accelerated computing with up to 21 TOPS provides the processing power necessary for modern neural networks to run in parallel and process data from multiple high-resolution sensors, making it a prerequisite for complete AI systems.

The Jetson Xavier NX Developer Kit is ready for mass production and works with any standard artificial intelligence library.

Powerful 21 TOPS AI Performance

The high-performance compute, and AI capabilities of embedded and edge systems are well served by the Jetson Xavier NX, which provides up to 21 TOPS. You will have access to the power of 384 NVIDIA CUDA Cores, 48 Tensor Cores, 6 Carmel ARM CPUs, and 2 NVIDIA Deep Learning Accelerators (NVDLA). These capabilities, along with the Jetson Xavier NX's over 51GB/s of memory bandwidth, video encoding, and decoding, make it the ideal platform for running multiple modern neural networks in parallel and processing high-resolution data from various sensors in realtime.

Figure 2-52　Jetson Xavier NX

Chapter 2 Introduction to Embedded Robotics

Table 2-7 Jetson Xavier NX Specifications

GPU	NVIDIA Volta architecture with 384 NVIDIA CUDA cores and 48 Tensor cores
CPU	6-core NVIDIA Carmel ARM® v8.2 64-bit CPU 6 MB L2 + 4 MB L3
Memory	8 GB 128-bit LPDDR4 × 51.2 GB/s
Storage	microSD (Card not included)
Power	10 W ∣ 15 W
PCIe	1 × 1 (PCIe Gen3) + 1 × 4 (PCIe Gen4), total 144 GT/s *
Camera	2 × MIPI CSI-2 D-PHY lanes
Video Encode	2 × 4Kp30 ∣ 6 × 1080p 60 ∣ 14 × 1080p30 (H.265/H.264)
Video Decode	2 × 4Kp60 ∣ 4 × 4Kp30 ∣ 12 × 1080p60 ∣ 32 × 1080p30 (H.265) 2 × 4Kp30 ∣ 6 × 1080p60 ∣ 16 × 1080p30 (H.264)
Display	HDMI and DP
DLA	2 × NVDLA Engines
Vision Accelerator	7-Way VLIW Vision Processor
Connectivity	Gigabit Ethernet, M.2 Key E (WiFi/BT included), M.2 Key M (NVMe)
USB	4 × USB 3.1, USB 2.0 Micro-B
Others	GPIO, I2C, I2S, SPI, UART
Mechanical	103 mm × 90.5 mm × 34 mm

Nvidia AGX Xavier

Next-generation robots rely heavily on visual odometry, sensor fusion, localization and mapping, obstacle detection, and path planning algorithms; Jetson AGX Xavier supports them all as a robust AI computer explicitly built for autonomous machines.

Get GPU workstation-class performance in a space-saving 100 × 87 mm form factor, with up to 32 TeraOPS (TOPS) peak compute and 750 Gbps of high-speed I/O.

Figure 2-53　Jetson AGX Xavier

Table 2-8　Jetson AGX Xavier Technical Specifications

GPU	512-core Volta GPU with Tensor Cores 11 TFLOPS (FP16) 22 TOPS (INT8)
CPU	8-core ARM v8.2 64-bit CPU, 8MB L2 + 4MB L3
Memory	32 GB 256-Bit LPDDR4× ∣ 137 GB/s
Storage	32 GB eMMC 5.1
DL Accelerator	(2×) NVDLA Engines 5 TFLOPS (FP16), 10 TOPS (INT8)
Vision Accelerator	7-way VLIW Vision Processor
Encoder/Decoder	(2×) 4Kp60 ∣ HEVC/(2×) 4Kp60 ∣ 12-Bit Support
Size	105 mm × 105 mm × 65 mm
Deployment	Module (Jetson AGX Xavier)

Table 2-9　Jetson AGX Xavier Development Board Specifications

PCIe X16	×8 PCIe Gen4/ ×8 SLVS-EC
RJ45	Gigabit Ethernet
USB-C	2× USB 3.1, DP (Optional), PD (Optional) Close-System Debug and Flashing Support on 1 Port
Camera Connector	(16×) CSI-2 Lanes

Chapter 2 Introduction to Embedded Robotics

(Continued)

M.2 Key M	NVMe
M.2 Key E	PCIe ×1 + USB 2.0 + UART (for Wi-Fi/LTE)/I2S/PCM
40-Pin Header	UART + SPI + CAN + I2C + I2S + DMIC + GPIOs
HD Audio Header	High-Definition Audio
eSATAp + USB3.0 Type A	SATA Through PCIe ×1 Bridge (PD + Data for 2.5-inch SATA) + USB 3.0
HDMI Type A	HDMI 2.0
uSD/UFS Card Socket	SD/UFS

Sensors

How do living beings obtain information regarding objects in their living environments? Their biological sensors have been developed over millions of years, and some perceive environmental data through their optical, acoustic, or ultrasonic (in-bath) sensors.

Plants could be more exceptional in receiving environmental data. There are also some researchers to connect and use plant sensors.

Plants also give off electrical signals like humans do when they move. By analysing and classifying these signals, it's possible to tell what stimulus may have been the cause. So, there's a signal for, say, acid, a chemical, fire, or many hundreds of things. Once you can read the signals accurately and tell them apart, you can use plants as biosensors. The trees come into the network.

Figure 2-54 Electric Signal

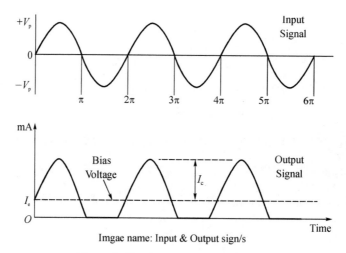
Imgae name: Input & Output sign/s

Figure 2-55 Input & Output Signals

Employing plants may be the beginning of developing some early building blocks. Just as humans are likely to be increasingly connected to machines, creating one seamless sensing network, we may also connect plants in the future. Then the division between the natural world and the human world may begin to disappear.

What we learn from this is:

1—Sensors are part of living beings in our natural world;

2—By developing various sensors, we aimed to make the connection between us and the natural world stronger.

Sensor's Classifications

There are three types of sensors:

1—Direct and Complex Sensors

2—Active and Passive Sensors

3—Absolute and Relative Sensors

Direct Sensors

A direct sensor by using an appropriate physical effect converts a stimulus into an electrical signal or modifies an electrical signal.

An example of direct sensors: Light Dependent Resistor (LDR).

LDR is a variable resistor whose resistance changes with light intensity in the operating environment. It is constructed from a thin film of cadmium sulphide sandwiched between two metal electrodes.

LDR chrematistic can be seen in the following graph:

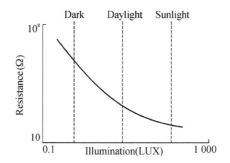

Figure 2-56 LDR Sensor Operation Graph

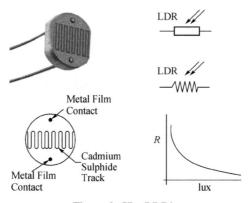

Figure 2-57 LDR's

Complex Sensors

A complex sensor in addition needs one or more transducers of energy before a direct sensor can be employed to generate an electrical output[17].

Example of complex sensors including:

Ultrasonic, Accelerometers, GPS, Precise Temperature sensors

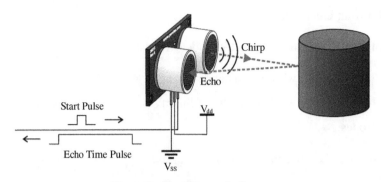

Figure 2-58 Ultrasonic Sensor

Ultrasonic or ping sensors are used in various devices, from measurement tools to car parking-assisted systems. Their operation is based on sending a chirp signal and waiting to receive the echo signal after hitting an obstacle—the time the signal travels and returns is used to calculate the distance.

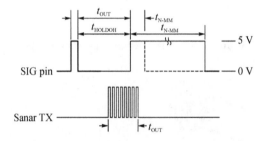

Figure 2-59 Ultrasonic Operation Principles

Passive Sensors

In response to a stimulus, the energy which comes from outside the sensor, a **passive sensor** converts this energy into an electric signal without any additional power source.

The examples are:

a thermocouple

a photodiode

a piezoelectric sensor

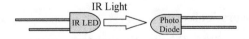

Figure 2-60 Photo Diode Sensor

Photodiodes are an example of passive sensors. Photodiodes are generally used where obstacle detection is required without precise distance.

Most passive sensors are direct sensors, as we defined them earlier.

Figure 2-61   IR Obstacle Sensor   Figure 2-62   IR Proximity Sensor

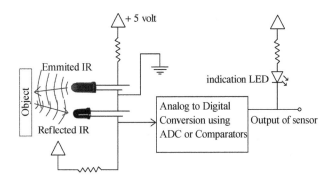

Figure 2-63   IR Transmitter and Receiver Working Principles

Active Sensors

In the active sensor, the excitation signal provides the active sensors with the external power necessary for operation.

The output signal is a modified version of the input signal that the sensor generates. Since the active sensors' properties can change in response to an external effect, and these changes are converted into electric signals, the active sensors are sometimes referred to as parametric sensors.

A thermistor is an example of an active sensor. In electronics, a thermosiphon, or thermistor, is a resistor that changes resistance depending on temperature.

A thermistor cannot produce an electric signal on its own. Still, when an electric current is passed through (excitation signal), the resulting changes in current and voltage across the thermistor can be used to calculate its resistance.

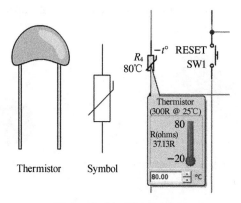

Figure 2-64　Thermistor

Assuming, as a first-order approximation, that the relationship between resistance and temperature is linear, then:

$\Delta R = k \Delta T$

Where ΔR, changes in resistance

And ΔT is changes in temperature

K, first order temperature coefficient of resistance

Thermistors are small, stable, long-lasting, and usually accurate to within $+/-.05\%$ to $+/-.02\%$. This feature makes thermistors superior to thermocouples and other devices that measure temperature.

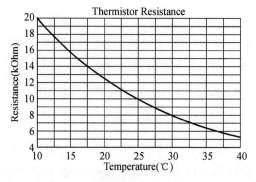

Figure 2-65　Resistance-temperature Characteristics of a Thermistor

The disadvantage is that, like typical semiconductors, they are non-linear, so this effect must be compensated for when building circuits. Also, unlike thermocouples, they cannot be used at very high temperatures.

Absolute Sensors

An absolute sensor detects a stimulus about an absolute physical scale independent of the measurement conditions, whereas a relative sensor produces a signal related to a particular case.

Thermistors also are considered absolute sensors.

Relative Sensors

Another viral temperature sensor—a thermocouple—is a 'relative sensor' and produces an electric voltage that is the function of a temperature gradient across the thermocouple wires. Thus, a thermocouple output signal cannot be related to any particular temperature without reference to a known baseline.

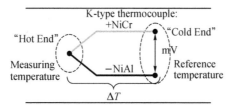

Figure 2-66   K-type Thermocouple

Transfer Function & Sensor Characteristics

Every sensor has a hypothetical or ideal output-stimulus relationship.

The output of a perfectly designed and constructed sensor made from ideal materials by ideal workers using ideal tools would always be an accurate reflection of the stimulus.

If 'S' is the electrical signal produced by sensor
And 's' is the stimulus, then

$$S = f(s)$$

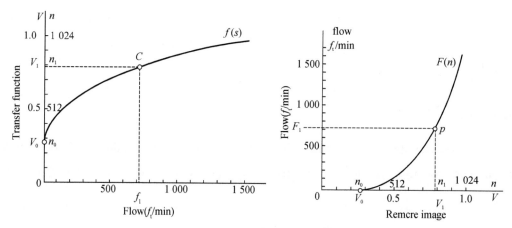

Figure 2-67   Transfer Function

Sometimes the transfer function is multi-dimensional, like in thermal radiation which the sensor output voltage not only depends on the temperature of the object but also on the temperature of the sensor's surface.

Another example is the Sharp IR Proximity sensor with the following transfer function:

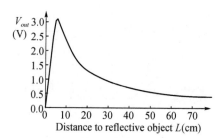

Figure 2-68   Sharp IR Proximity Sensor

Full Scale Input (FSI)

A dynamic range of stimuli a sensor may convert is called a span or a full scale input (FSI). It represents the highest possible input value that can be applied to the sensor without causing an unacceptably significant inaccuracy.

For the sensors with an extensive and nonlinear response characteristic, a dynamic range of the input stimuli is often expressed in decibels, which is a logarithmic measure of either power or force (voltage) ratios. Decibels do not measure absolute values but a ratio of values only.

Chapter 2 Introduction to Embedded Robotics

Full Scale Output (FSO)

Full Scale Output (FSO) is the algebraic differences between the output signals measured with maximum input stimulus and with minimum input stimulus applied.

Full Scale Output is the algebraic differences between the output signals measured with maximum input stimulus and with minimum input stimulus applied.

Full scale Output is the difference between the electrical output signals measured with the maximum input stimulus and the lowest input stimulus applied. This must include all deviations from the ideal transfer function. For instance, the FSO output in Figure 2-69 is represented by SFS.

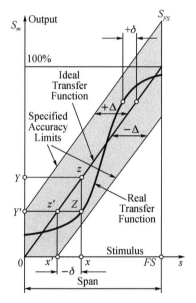

Figure 2-69 Sensor Full Scale Output

Accuracy

An essential characteristic of a sensor is accuracy which means inaccuracy. Inaccuracy is measured as the highest deviation of a value the sensor represents from the ideal or actual value at its input.

A real function rarely coincides with the ideal. Because of material variations, built, design errors, manufacturing tolerances, and other limitations, it is possible identical conditions.

Saturation

Every sensor has its operating limits. Even if it is considered linear, its output signal will no longer be responsive at some levels of the input stimuli. A further increase in stimuli does not produce a desirable output. It is said that the sensor exhibits a span-end nonlinearity or saturation.

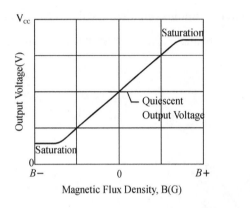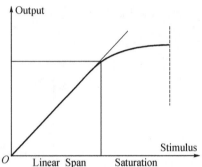

Figure 2-70   Saturation

Repeatability

Failing to represent the same value by a sensor under an identical condition is called a repeatability (reproducibility) error. Repeatability error is expressed as the maximum difference between output readings determined by two calibrating cycles unless otherwise specified. It is usually represented as % of FS:

$$\delta_r = \frac{\Delta}{FS} \times 100\%$$

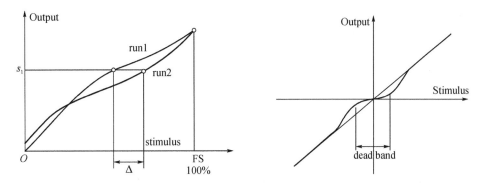

Figure 2-71   Error Produced by Repeatability

Dead Band

The dead band is the insensitivity of a sensor in a specific range of input signals. In that range, the output may remain near a particular value (often zero) over an entire dead band zone.

Resolution

Resolution describes the smallest increments of stimulus which can be sensed. When a stimulus continuously varies over the range, the output signals of some sensors will not be perfectly smooth, even under no-noise conditions.

Output Impedance

Knowing the output impedance is essential for a better sensor interface with the electronic circuit. Most popular sensors have an output impedance greater than several Mega Ohm.

Sensor Modules

Sensor modules are small PCB boards with a sensor and minimum required supporting electronic components to make the sensor to be use with modern development boards. They may work based on a 5-volt logic level for Arduino UNO and similar, or 3.3 volts for most development boards such as Arduino Nano BLE or Raspberry Pi.

If a sensor doesn't match our development's logic level, we may use a logic-level converter board to interface between the sensor and the development board. They usually are in 4, and 8-channel combinations.

Sensor Shields

Sensor shields give us more flexibility and more accessible connections between our sensors and Arduino boards as well as some extra facilities, mainly if we use sensor modules for quick prototyping.

Important: To connect or disconnect any device, including sensor shields, you must power off your Arduino first.

Figure 2-72　Arduino and Sensor Shield　　　Figure 2-73　Sensor Shield

Connecting to Arduino:

Check pin compatibility between your sensor module and the sensor shield if you use a sensor shield. To connect any sensor to your sensor shield, you must always follow the correct connections, even if the link of your sensor pins doesn't match the one on the sensor shield. Sometimes there are different positions on sensor modules or colours used in sensor cables.

Usually, sensor modules have:

VCC (or +): this is a positive power supply and uses a red jumper wire to connect it to VCC on Arduino or a sensor shield. It is crucial to know the operating voltage of the sensor (3.3 V or 5 V); otherwise, you may damage the module.

GND (or -): This is a negative power supply or ground and must be connected to GND on your Arduino or sensor shield. The colour code for this connection is black.

OUT: This is the sensor's output and must be connected to the desired Arduino digital or analogue I/O. You must know what type of output your sensor provides (digital/analogue).

DOUT: This digital output based on a voltage threshold above 3 volts returns HIGH or LOW. This pin can be connected to Arduino digital pins D0-D13, and you can choose any colour wire.

Some sensors have additional pins such as **AOUT**, or **EN**.

AOUT: This is an analogue pin to return a voltage level between 0-5 volts. This pin can be connected to Arduino analogue pins A0-A5 and use any colour wire.

Figure 2-74　Sensor Modules

Optical Sensors

There are many types of optical sensors for different projects. The most common are LDR, IR (Infra-Red Obstacle), colour, and Passive Infra-Red sensors. The goal of this section is to use more valuable examples, diagrams, and code snippets to make all the theoretical knowledge more tangible and, of course, to improve skills in building practical, intelligent robots.

Light Sensor or LDR (Light Dependent Resistor)

Working principal?

Further to our discussion about LDR sensors in 'direct sensors type', the circuit bellows work similarly to a voltage divider but with LDR (R2) as a variable resistor. LDR (R2) resistance will change with the light intensity, changing the voltage at A0.

Example:

An LDR has a resistance of 15 ohms at a very high light level. What value of a protection resistor is needed if a current of no more than ten mA flows when the supply voltage is 9.0 volts?

Current through LDR = 10 mA = 0.01 A

Voltage across LDR = 0.01 A x 15 Ω = 0.15 V

Resistance = 8.85 V/0.01 A = 885 Ω

The general formula for protection resistance and output voltage can be calculated using this formula:

$$V_{out} = V_s X \frac{R_2}{R_1 + R_2}$$

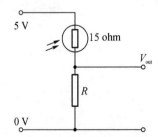

Figure 2-75　LDR Circuit Diagram

Chapter 2 Introduction to Embedded Robotics

The top of the Potential Divider is 5 V; the bottom is at 0V; the middle (connected to A0) is some value between 5 V and 0 V that varies as the LDR resistance varies. Remember, the LDR resistance varies with Light, so the Voltage at A0 will too.

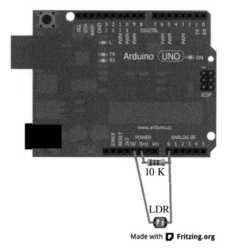

Figure 2-76 LDR Connection to Arduino

```
/*
 * LDR Sensor module
 * Robotics and AI Labs
 * 2015 - 2022
 * HuaiYin Institute of Technology
 */

int LDR = 0;

void setup(){
  Serial.begin(9600);
}

void loop(){
  int Reading = analogRead(LDR);
  Serial.print("Reading:");
  Serial.println(Reading);
  delay(250);
}
```

Figure 2-77 LDR Example Program

Figure 2-78　LDR Program Output

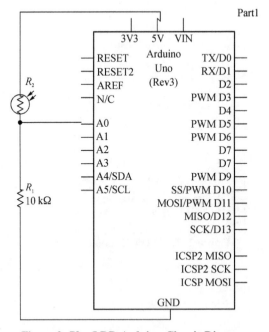

Figure 2-79　LDR-Arduino Circuit Diagram

Chapter 2 Introduction to Embedded Robotics

If you want to be very precise and technical, then you can work out the voltage on A0 as:

$$V_{A0} = 5 \times R_1 / (R_1 + R_2)$$

where V_{A0} is the voltage at A0 pin, R_2 is the top resistor value, R_1 is the bottom resistor value;

e.g. $R_1 = 10\ k$, $R_2 = 5\ k \Rightarrow V_{A0} = 5 \times 10000/(10000 + 5000) = 5 \times 10/15 = 3.33\ V$

The LDR has a high value when no light is present. The resistance value of the LDR depends on the type; in this case, it's about 10 k. As the light level increases, the resistance drops, which increases the current (by Ohm's Law), which increases the voltage at A0 (V_{A0}).

$$I = 5/(R_1 + R_2)$$

Now, the voltage across the resistor is applied to A0. Again, by Ohm's Law that is:

$$V_{A0} = I \times R_1$$

Substituting the equation for I back in to this equation we get:

$$V_{A0} = 5 \times R_1 / (R_1 + R_2)$$

For prototyping, sometimes we use an LDR sensor module, which already has a built-in resistor, and a digital out circuit to detect a light threshold adjustable with the onboard potentiometer.

LDR Sensor Module

There are many different LDR sensor modules available. Some only offer digital output as two conditions of 'light' or 'dark' set by an onboard potentiometer. Some provide both digital and analogue output to receive a magnitude correspondence to environmental light.

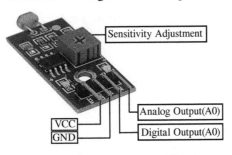

Figure 2-80　LDR Sensor Module

Connecting light sensor to sensor shield:

The sensor shield is designed to connect to sensors with either analogue or digital output, but not both. You may not be able to use three colour sensor wires for analogue output. Use a single wire from your sensor to any analogue pins to your sensor shield or Arduino board[18].

This sensor can detect light intensity (brightness) with two different outputs (DOUT—digital work to return a fixed threshold, and AOUT—analogue output for voltage differences).

TECHNICAL PARAMETERS

- Size: 28 mm × 24 mm
- Voltage: 3.3 V, 5 V
- Output: Dual output (analog, digital)
- Probe: CSD light sensors
- Detection: Visible light
- Detection angle: 60°
- Detection: Dividing feedback
- Platform: Arduino
- DOUT: When the threshold is reached by setting the threshold digital outputs
- AOUT: Analog voltage output

Passive Infra-Red (PIR)

PIR (Passive Infra-Red) detectors can detect changes (move) in a heat body such as a human. A person's movement in front of the sensor can trigger the sensor.

Mini PIR module:

This mini passive infrared sensor is especially designed for prototyping with development boards. It can sense the movement of people, animals, or other objects. It features high sensitivity and instant reaction, commonly used in burglar alarms and automatically-activated lighting systems.

Chapter 2　Introduction to Embedded Robotics

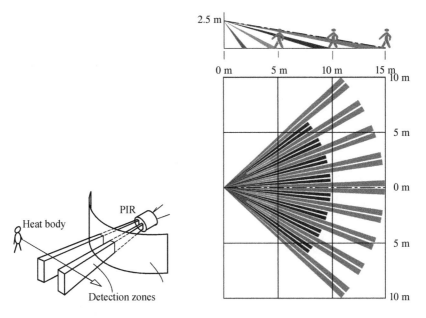

Figure 2-81　Standard PIR Working Principle

Figure 2-82　Mini PIR Module

Example code using interrupt:

The example here will output "PIR Activated" on the serial monitor as soon as the PIR is activated.

```
const int PIR _PIN = 2

void setup( ) {
    pinMode( PIR_PIN, INPUT_PULLUP);
    attachInterrupt[ digitalPinToInterrupt(2), pir_action, FALLING];
```

```
        Serial.begin(9600);
        delay(20000);
    }

    void loop() {
    }

    void pir_action() {
        Serial.println("PIR Activated");
    }
```

Infra-Red Obstacle Sensor

If an object is in front of the sensor, the output is **LOW**. Otherwise, it is **HIGH**. Two potentiometers can adjust the sensitivity and the detection distance.

Figure 2-83   Infrared Obstacle Sensor

The detection range is 2-40 cm but varies with the size and colour of objects. Also, the lite intensity can change the detection sensitivity. The content of 2-40 cm is calculated in front of a white background, such as a white wall.

Input voltage between 3-5 volt

A frequency adjustment potentiometer adjusts the infrared emission frequency; the maximum 38 kHz frequency is the most sensitive.

Below are the connection pins for Arduino and the sensor:

Chapter 2 Introduction to Embedded Robotics

Table 2-10 Infrared Obstacle Sensor Pinouts

Sensor	Arduino
GND	GND
+	5v
Out	PIN 9
EN	Not used

Temperature Sensors

Temperature sensors play an important role in nearly every industry. There are four common temperature sensors used in different projects:

- Resistance Temperature Detectors (RTDs)
- Thermistors (Negative Temperature Coefficient, NTC)
- Thermocouples
- Semiconductor-Based Sensors

Thermistor-Based Temperature Measurement

While many temperature sensor modules are available, sometimes, for different reasons, we may need to use a thermistor directly connected to a microcontroller. [19]

Figure 2-84 Temperature sensors

Here in this example, Arduino Mini is selected as a small-range microcontroller.

To measure the voltage, we connect the thermistor in series to another fixed resistor R in a voltage divider circuit. The variable thermistor resistance is labelled as R0.

We select a thermistor with a resistance of 10 kΩ at 25 ℃ and a fixed resistance of 10 kΩ [20].

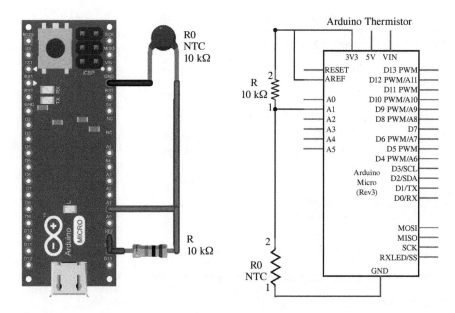

Figure 2-85　Connecting a Thermistor to Arduino Micro (Voltage Divider Circuit)

Vcc of the voltage divider circuit is connected to 3V on Arduino Micro, which provides a 3.3-volt supply generated by the onboard regulator with a maximum current draw of 50 mA.

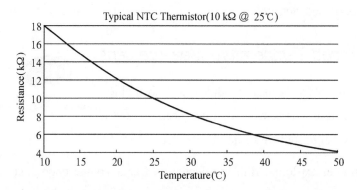

Figure 2-86　NTC Thermistor Changes in Resistance

The 3 V pin is connected to the AREF pin because we need to change the upper reference of the analogue input range.

If the output voltage is V_0, the power supply V_{cc}, the variable thermistor resistance R_0 and the fixed resistance R, the output voltage is given by:

Then

Chapter 2 Introduction to Embedded Robotics

$$V_0 = (V_{cc} \cdot R_0)/(R_0 + R)$$

The output voltage is connected to the Arduino analogue input pin A_1. Arduino Micro provides a 10-bit ADC Analog to Digital Converter, which means that the output voltage is converted in a number between 0 and 1023.

Then, if A_1 is the ADC value measured by Arduino Micro, then the output voltage is given by:

$$V_0 = A_1 \cdot V_{cc}/1023$$

By combining the previous equations, we have:

$$V_{cc} \cdot R_0/(R_0 + R) = A_1 \cdot V_{cc}/1023 \Rightarrow R_0/(R_0 + R) = A1/1023$$

R_0, the thermistor's resistance is independent of the supply voltage V_{cc}.

For temperature measurement, we need the thermistor's variable resistance R_0. Using the previous equation:

$$R_0 = A_1 \cdot R/(1023 - A1)$$

The resistance measurement depends on the ADC of Arduino's A1 input, the fixed resistor R in the voltage divider, and the ADC resolution of Arduino, which is 1023.

$$V_{out} = IR_2$$

$$V_{out} = \frac{V}{R_1 + R_2} X R_2$$

$$V_{out} = V_s X \left(\frac{V}{R_1 + R_2}\right)$$

From the above; the output voltage is the same fraction of input voltage as R_2 is the fraction of total resistance.

Put them all together:

$$V_0 = \frac{(V_{cc} \cdot R_0)}{R_0 + R}$$

$$V_0 = A_1 \cdot V_{cc}/1023$$

$$V_{cc} \cdot R_0/(R_0 + R) = A_1 \cdot V_{cc}/1023 \Rightarrow R_0/(R_0 + R) = A1/1023$$

$$R_0 = A_1 \cdot R/(1023 - A_1)$$

To improve the output's accuracy, the Arduino board's ADC has a 10-bit ADC resolution. We use multiple sampling and take the average. Obtaining an average of over ten samples to smooth the ADC data can provide better accuracy.

To transfer resistance to temperature, we can use the Steinhart-Hart equation[19] (aka B or β parameter equation), which is a good approximation of the resistance to temperature relation:

$$1/T = 1/T_0 + 1/B \cdot \ln(R/R_0)$$

where R is the thermistor resistance at the generic temperature T, R_0 is the resistance at T_0 = 25 ℃ and B is a parameter depending on the thermistor. The B value is typically between 3000-4000.

Arduino program:

```
#define THERMISTORPIN A1              // Thermistor pin
#define THERMISTORNOMINAL 10000       // resistance at 25 degrees C
#define TEMPERATURENOMINAL 25         // temp. for nominal resistance (almost always 25 C)
#define NUMSAMPLES 10                 // how many samples to take and average
#define BCOEFFICIENT 3950             // The beta coefficient of the thermistor (usually 3000-4000)
#define SERIESRESISTOR 10000          // the value of the 'other' resistor

void setup () {

}

int getTemperature(void){
    int i;
    float average;
    int samples[NUMSAMPLES];
    float thermistorResistance;
    int Temperature;
    for (i=0; i< NUMSAMPLES; i++) {   // acquire N samples
        samples[i] = analogRead(THERMISTORPIN);
        delay(10);
    }
    average = 0;                      // average all the samples out
    for (i=0; i< NUMSAMPLES; i++) {
        average += samples[i];
    }
    average /= NUMSAMPLES;
    thermistorResistance = average * SERIESRESISTOR / (1023 - average);   // convert the value to resistance
    float steinhart;
    steinhart = thermistorResistance / THERMISTORNOMINAL; // (R/Ro)
    steinhart = log(steinhart); // ln(R/Ro)
    steinhart /= BCOEFFICIENT; // 1/B * ln(R/Ro)
    steinhart += 1.0 / (TEMPERATURENOMINAL + 273.15); // + (1/To)
    steinhart = 1.0 / steinhart; // Invert
    steinhart -= 273.15; // convert to C
    // decimal value
    Temperature = steinhart * 10;
    return(Temperature);
}

Void loop() {
getTemprature() {
}
```

Capacitive Touch Sensors

Capacitive touch sensor principle is based on capacitance characteristics. The simplest form of capacitor can be made with two conductors separated by an insulator. Metal plates can be considered as conductors.

The formula of capacitance is shown below.

$$C = \varepsilon_0 \cdot \varepsilon_r \cdot A/D$$

Where:

ε_0 is the permittivity of free space;

ε_r is relative permittivity or dielectric constant;

A is area of the plates and D is the distance between them.

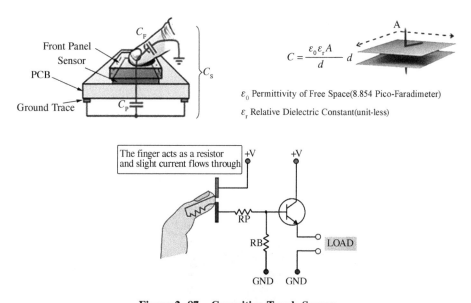

Figure 2-87 Capacitive Touch Sensor

Touch sensors are also called tactile and sensitive to touch, force or pressure. They are one of the most straightforward and popular sensors. The working of a touch sensor is similar to that of a simple switch. When there is contact with the surface of the touch sensor, the circuit is closed inside the sensor, and there is a flow of current. When the contact is released, the circuit is opened, and no current flows.

In capacitive touch sensors, the electrode represents one of the plates of the capacitor. Two objects represent the second plate:

One is the environment of the sensor electrode which forms parasitic capacitor C0, and the other is a conductive object like a human finger which forms touch capacitor CT. The sensor electrode is connected to a measurement circuit, and the capacitance is measured periodically. The output capacitance will increase if a conductive object touches or approaches the sensor electrode. The measurement circuit will detect the capacitance change and convert it into a trigger signal.

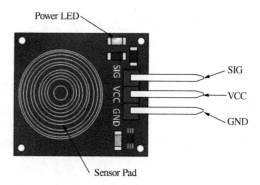

Figure 2-88 Capacitive Touch Sensor

Benefits of capacitive touch sensors are including:

- Each touch sensor requires only one wire connected to it.
- Can be concealed under any non-metallic material.
- Can be easily used in place of a button.
- Can detect a hand from a few inches away, if required.
- Very inexpensive.

The diagram below demonstrates a connection of a capacitive touch sensor to the Arduino UNO development board.

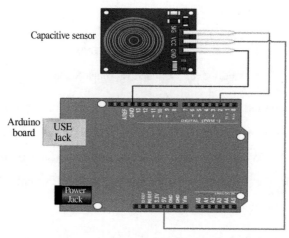

Figure 2-89 Capacitive Touch Sensor Circuit Diagram

Chapter 2 Introduction to Embedded Robotics

An example Arduino program for capacitive touch sensor:

Figure 2-90 Arduino Example for Capacitive Touch Sensor

Proxy

Proximity sensors are devices to detect the presence of objects near the sensor operation range. The important key point in proximity sensors is that they are a non-contact type of sensor, with various proximity sensors for different applications.

Depending on the object detection method, there are four widely-used types of proximity sensors and some newer high-end designs[21]:

- Inductive Proximity Sensors
- Capacitive Proximity Sensors
- Ultrasonic Proximity Sensors
- IR Proximity Sensors
- High-end Proximity Sensors

Capacitive Sensors and Inductive Proximity Sensors

Industrial capacitive sensors have a more accurate distance of detection and are employed in many fields, including production lines.

Inductive proximity sensors are used for the non-contact detection of metallic objects. In contrast, capacitive proximity sensors are more common for non-contact detection of metallic and non-metallic objects such as liquid, plastic, paper and more.

Figure 2-91   Capacitive Sensor

The figure below is a very simple Arduino code to demonstrate operation of an inductive proximity sensor.

```
/*
 * Inductive Proximity sensor
 * Robotics and AI Labs
 * 2015 - 2022
 * HuaiYin Institute of Technology
 */
#define s1 2
void setup() {
    pinMode(s1, INPUT);
    Serial.begin(9600);
}

void loop() {
    int sensorValue = digitalRead(s1);
    if(sensorValue==HIGH){
        Serial.println("no detection!");
        delay(500);
    }
    else{
        Serial.println("Detected!");
        delay(500);
    }
}
```

Chapter 2 Introduction to Embedded Robotics

Hall Effect Sensor

KY-0003

The KY-003 is a Hall effect sensor. If no magnetic field is present, the signal line of the sensor is HIGH. If a magnetic field is presented to the sensor, the signal line goes LOW, and at the same time, the LED on the sensor lights up.

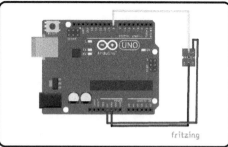

```
1.  int Led = 13 ; // Arduino built in LED
2.  int SENSOR = 8 ; // define the Hall magnetic sensor
3.  int val ; // define numeric variables
4.
5.  void setup ()
6.  {
7.     Serial.begin(9600);
8.     pinMode (Led, OUTPUT) ;    // define LED as output
9.     pinMode (SENSOR, INPUT) ;  // Hall magnetic sensor line as input
10. }
11.
12. void loop ()
13. {
14.    val = digitalRead (SENSOR) ; // read sensor line
15.    if (val == LOW) // when a magnetic field detected, LED lights up
16.         {
17.            digitalWrite (Led, HIGH);
18.            Serial.println("Magnetic field detected");
19.         }
20.    else
21.         {
22.            digitalWrite (Led, LOW);
23.            Serial.println("No magnetic field detected");
24.         }
25.    delay(1000);
26. }
```

RFID

RC-522

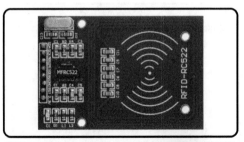

FID is an acronym for "radio-frequency identification". It refers to a technology whereby digital data encoded in RFID tags or smart labels (defined below) are captured by a reader via radio waves. It is widely used in door access control.

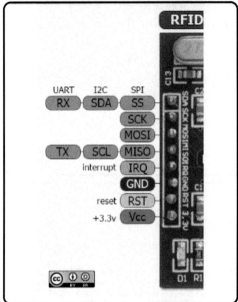

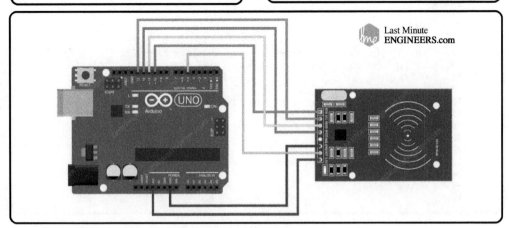

Chapter 2   Introduction to Embedded Robotics

RFID Library example snippet at: https://github.com/miguelbalboa/rfid

```
1.  #include <SPI.h>
2.  #include <MFRC522.h>
3.  #include <Ethernet.h>
4.  #define RST_PIN         9          // Configurable, see typical pin layout
5.  #define SS_PIN          10         // Configurable, see typical pin layout
6.  MFRC522 mfrc522(SS_PIN, RST_PIN);  // Create MFRC522 instance
7.  byte mac[] = {
8.    0xDE, 0xAD, 0xBE, 0xEF, 0xFE, 0xED };
9.  // Enter the IP address for Arduino, as mentioned we will use 192.168.0.16
10. // Be careful to use , instetead of . when you enter the address here
11. IPAddress ip(192,168,1,193);
12. char server[] = "192.168.1.208";
13. // Initialize the Ethernet server library
14. EthernetClient client;
15.
16. void setup() {
17.   Serial.begin(9600); // Initialize serial communications with the PC
18.   while (!Serial);
19.   SPI.begin(); // Init SPI bus
20.   mfrc522.PCD_Init(); // Init MFRC522
21.   delay(4);
22.   mfrc522.PCD_DumpVersionToSerial(); details
23.   Serial.println(F("Scan PICC to see UID, SAK, type, and data blocks..."));
24. }
25.
26. void loop() {
27.   when idle.
28.   if ( ! mfrc522.PICC_IsNewCardPresent()) {
29.   return;
30.   }
31.
32.   // Select one of the cards
33.   if ( ! mfrc522.PICC_ReadCardSerial()) {
34.   return;
35.   }
36.
37.   // Dump debug info about the card; PICC_HaltA() is automatically called
38.   mfrc522.PICC_DumpToSerial(&(mfrc522.uid));
39. }
```

Gas Sensor

MQ-X

The gas-sensitive material used in MQ-2 gas sensors is SnO2, with lower conductivity in clean air. With the increase of the concentration of combustible gas around the sensor, the conductivity of the sensor increases. A simple circuit can be used to convert the electrical conductivity into the output signal corresponding to the gas concentration. MQ-2 gas sensor is susceptible to LPG, propane and hydrogen; it can be used for gas leakage monitoring devices in residential and industrial.

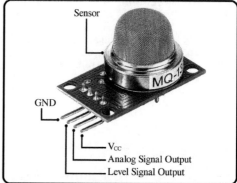

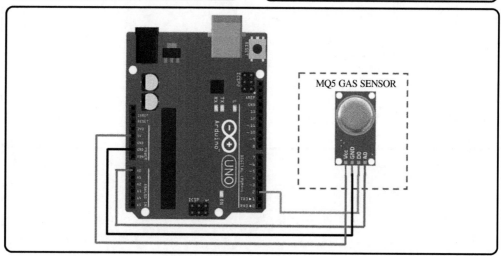

Chapter 2 Introduction to Embedded Robotics

```
1.  /*
2.   * Magnetic Reed Sensor
3.   * Robotics and AI Labs
4.   * 2015 - 2022
5.   * HuaiYin Institute of Technology
6.   */
7.  int Led_pin = 12 ;      // initializing the pin 12 as led pin
8.  int Sensor_pin = 2 ;    // initializing the pin 2 sensor pin
9.  int Value ;             // initializing a variable to store  sensor
10.
11. void setup ( ) {
12.    pinMode ( Led_pin, OUTPUT ) ;
13.    pinMode ( Sensor_pin, INPUT ) ;
14. }
15.
16. void loop ( )
17. SunFounder {
18.    Value = digitalRead ( Sensor_pin ) ;
19.    if (Value == HIGH ) {
20.       digitalWrite ( Led_pin, HIGH ) ;
21.    }
```

There are different sensors sensitive to different chemical gas.

MQ-2 Combustible gas, Smoke	MQ-3 Alcohol	MQ-4 Methane, Propane, Butane
MQ-5 Methane, Propane, Butane	MQ-6 Liquefied Petroleum Butane, Propane, LPG	MQ-7 Carbon Monoxide
MQ-8 Hydrogen	MQ-9 Carbon Monoxide, Methane	MQ-135 Ammonia Sulfide, Benzene Vapor

Magnetic Reed Switch

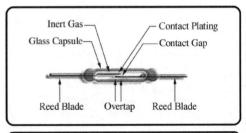

Reed switch is a virtual sensing device widely used in security (including burglar alarms, access control, home automation, intelligent control and more). It has many other names, such as magnetic contact or door sensor. The reed switch becomes the most reliable and cost-effective intrusion sensor in the burglary detection area.

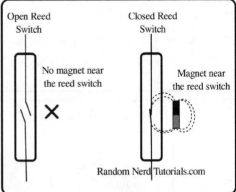

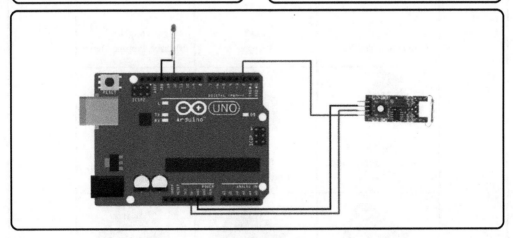

```
1.  /*
2.   * Magnetic Reed Sensor
3.   * Robotics and AI Labs
4.   * 2015 - 2022
5.   * HuaiYin Institute of Technology
6.   */
7.  int Led_pin = 12 ;      // initializing the pin 12 as led pin
8.  int Sensor_pin = 2 ;   // initializing the pin 2 sensor pin
9.  int Value ;             // initializing a variable to store  sensor
10.
11. void setup ( ) {
12.   pinMode ( Led_pin, OUTPUT ) ;
13.   pinMode ( Sensor_pin, INPUT ) ;
14. }
15.
16. void loop ( )
17. SunFounder {
18.   Value = digitalRead ( Sensor_pin ) ;
19.   if (Value == HIGH ) {
20.     digitalWrite ( Led_pin, HIGH ) ;
21.   }
```

Ultrasonic

HC-SR04

The HC-SR04 is an inexpensive ultra-sonic range finder suitable for education or prototyping. Many other models have more reliability and accuracy for a higher price.

It detects the distance of the closest object in front of the sensor (from 2 cm up to 400 cm). It works by sending out a burst of ultrasound and listening for the echo when it bounces off an object.

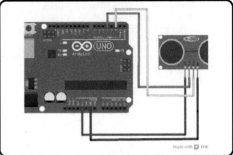

```
/*
*HC-SR04 Ultrasonic Range sensor
* Robotics and AI Labs
* 2015 - 2022
* HuaiYin Institute of Technology
*/
long tiempo;
int disparador = 7;   // triger
int entrada = 8;      // echo
float distancia;

void setup(){
    pinMode(disparador, OUTPUT);
    pinMode(entrada, INPUT);
    Serial.begin(9600);
}

void loop(){
    digitalWrite(disparador, HIGH);
    delayMicroseconds(10);
    digitalWrite(disparador, LOW);
    tiempo = (pulseIn(entrada, HIGH)/2); /
    distancia = float(tiempo * 0.0343);
    Serial.println(distancia);
    delay(1000);
}
```

Chapter 2 Introduction to Embedded Robotics

Various ultrasonic sensors are available with different ranges, waterproof, more accurate, more reliable, serial bus, RS485, CAN bus, half duplex, and full duplex; some return calculated centimetre distance. However, all follow the same principle.

Perhaps the Arduino NewPing library is more convenient and accurate; however, ultrasonic sensors are still open to unwanted reflections, returning the wrong signal and distance. Using averaging three good signals should reduce errors substantially, which can be found in Appendix B, Dealing with errors.

```
/*
 * HC-SR04 Ultrasonic Range sensor
 * Using 'NewPing' library
 * Robotics and AI Labs
 * 2015 - 2022
 * HuaiYin Institute of Technology
 */
#include <NewPing.h>
 #define TRIGGER_PIN 11
#define ECHO_PIN 12
#define MAX_DISTANCE 200

NewPing sonar(TRIGGER_PIN, ECHO_PIN, MAX_DISTANCE);

void setup() {
    Serial.begin(9600);
}

void loop() {
    delay(50);
    unsigned int uS = sonar.ping_cm();
    Serial.print(uS);
    Serial.println("cm");
}

```

Ultrasonic II

MaxSonar EZ1

MaxSonar EZ1 provides three ways of reading the sensor's output: Analog, PWM, and Serial.

The Analog Interface is not the most accurate but is the easiest way to communicate with this sensor; there is a directly proportional between Distance and Output voltage.

Outputs analogue voltage with a scaling factor of (Vcc/512) per inch.

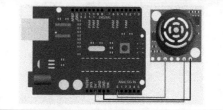

```
/*
 * MaxSonar EZ1
 * Robotics and AI Labs
 * 2015 - 2022
 * HuaiYin Institute of Technology
 */
float Inch=0.00;
float cm=0.00;
int SonarPin=A0;
int sensorValue;

void setup() {
  pinMode(SonarPin,INPUT);
  Serial.begin(9600);
}

void loop() {
  sensorValue=analogRead(SonarPin);
  delay(50);
  Inch= (sensorValue*0.497);
  cm=Inch*2.54;
  Serial.println(sensorValue);
  Serial.print(Inch);
  Serial.println("inch");
  Serial.print(cm);
  Serial.println("cm");
  delay(100);
}
```

Chapter 2 Introduction to Embedded Robotics

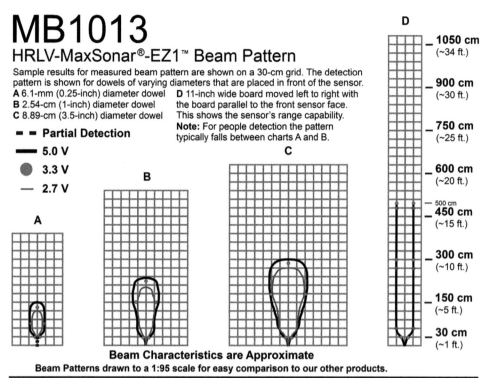

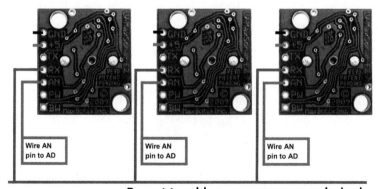

MaxSonar sensors can be cascaded in order to connect them to the serial bus.

Motors

It is essential to understand how different motors operate and can be controlled. In general, we will discuss three different types of motors, DC motors, Servo motors, and Stepper motors. Different projects require a different type of motors. Our first goal is to investigate different motors and their advantages and disadvantages. Then look at some examples of using different motors for different projects. For instance, wheels usually are best with DC motors for speed, strength, and cost. Arms and more precise movement applications often use servo motors.

DC Motors

Two-wire (power & ground), continuous rotation motors are known as DC (Direct Current) motors. A DC motor will start spinning when power is applied and continue spinning until power is cut off. Examples of DC motors that operate at high RPMs include computer cooling fans and radio-controlled car wheels.

Figure 2-92　A DC Motor

Chapter 2 Introduction to Embedded Robotics

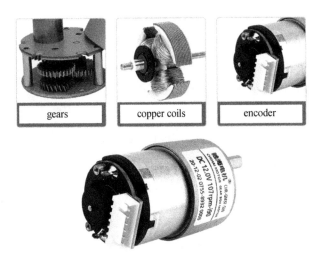

Figure 2-93 A Geared DC Motor with Encoder

Pulse width modulation (PWM), a method of quickly pulsing the power on and off, regulates the speed of DC motors. The speed of the motor is determined by the amount of time spent cycling the on/off ratio; for example, if the power is cycled at 50% (half on, half off), the motor will spin at half the speed of 100%. (fully on). Each pulse is so quick that the motor runs without interruption.

Each pulse is so rapid that the motor appears to be continuously spinning with no stuttering!

Servo Motors

A DC motor, a gearing set, a control circuit, and a position sensor typically (usually a potentiometer) make up a servo motor. Servo motors typically have three wires, allowing for more precise position control than standard DC motors (power, ground & control).

The servo control circuit controls the draw to drive the motor while power is continuously applied to servo motors.

Servo motors are designed to perform specific tasks where position accuracy is essential, like controlling a robot arm to perform a precise motion. Usually, servo motors do not rotate freely like a standard DC motor, and instead, the rotation angle is limited to 180 degrees (or so) back and forth. Servo motors receive a control signal representing an output position and apply power to the DC motor until the shaft turns to the correct position, determined by the position sensor. A PWM signal is used for the control signal of servo motors. However, unlike DC motors, the positive pulse's duration is used to determine the position, rather than speed, of the servo shaft. A neutral pulse value dependent on the servo (usually around 1.5ms) keeps the shaft in the centre position.

Figure 2-94　Dynamixel Servo Motor

Increasing the PWM pulse value will push the servo turn clockwise; a shorter pulse will turn the shaft anticlockwise. The PWM servo control pulse is usually repeated every 20 milliseconds, and moves the servo to the expected position, even if that means staying in the same position. When a control signal commands the servo to move, it will move to the position and hold that position, even if external force pushes against it. The servo will resist moving out of that position, with the maximum amount of resistive force the servo can exert is the torque rating of that servo.

Stepper Motors

A stepper motor can be considered a servo motor that uses a different moving method. As discussed earlier, a servo motor contains a continuous rotation DC motor and a controller circuit; stepper motors utilize multiple toothed electromagnets arranged around a central gear to define the position. An external control circuit or a microcontroller must energize each electromagnet and make the motor shaft turn. When one electromagnet is powered, it attracts the gear's teeth and aligns them, slightly offset from the next electromagnet. When the first one is switched off, the next one is switched on; the gear rotates slightly to align with the next one, and so on around the circle, with each electromagnet around the gear energizing and de-energizing in turn to create rotation. Each rotation from one electromagnet to the next is called a "step". Thus, precise pre-defined step angles can turn the motor through a full 360 degree rotation.

Chapter 2 Introduction to Embedded Robotics

Figure 2-95 Low Power Stepper Motor

Figure 2-96 Stepper Motor 42

Stepper motors are available in two types; unipolar and bipolar. Bipolar stepper motors, usually having four/eight leads, are the most vital type. In bipolar, there are two sets of electromagnetic coils internally, and stepping is achieved by changing the direction of current within those coils. Unipolar motors also have two coils but with a centre tap and can be identifiable by having 5, 6, or 8 wires. In unipolar motors, electronics are more straightforward as they can step without reversing the current direction in the coils. However, because only half of each coil at a time is used to energise, they have less torque power. The stepper motor holds torque without needing the motor to be powered, and, provided that the motor is used within its limits, positioning errors do not occur since stepper motors have physically pre-defined stations.

Summary

DC Motors are fast, continuous rotation motors—used for anything that needs to spin at a high RPM, e. g. car wheels and fans.

Servo Motors are fast, high torque, accurate rotation within a limited angle—generally a high-performance alternative to stepper motors, but the more complicated setup with PWM tuning. They are suited for robotic arms/legs or rudder control.

Stepper Motors are slow, precise rotation, and easy to set up & control—an advantage over servo motors in positional control.

Where servos require a feedback mechanism and support circuitry to drive positioning, a stepper motor has positional control via its nature of rotation by fractional increments—suited for 3D printers and similar devices where the position is fundamental.

Hardware Control

PID Control

PID (Proportional, Derivative and Integral) is one of the most common control schemes. Most industrial control loops use some flavour of PID control.

PID controllers are used wherever there is a need to control a physical quantity and to make it equal to a specified value: for example, Cruise controllers in cars, Robots, Temperature regulators, Voltage regulators, and more.

PID control continuously calculates an error value $e(t)$ as the difference between a required setpoint $SP = r(t)$ and a variable to measure the process

$$PV = y(t) : e(t) = r(t) - y(t)$$

And applies a correction based on proportional, integral, and derivative terms.

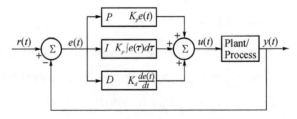

Figure 2-97　PID Control[22]

The PID control function therefore is:

$$\mu(t) = K_p e(t) + K_i \int_0^t e(\tau)\,d\tau + K_d \frac{de(t)}{dt}$$

Where K_p, K_i, and K_d all are non-negative coefficients for the proportional, integral, and derivative terms respectively.

PID control is an essential mechanism where moving accurately to a set point is required[22].

A position that a robot arm is going to move to,

A direction that a servo motor to turn to,

A path that a car has to go though,

An illumination a light has to reach to,

A distance from an object has to be taken.

For instance, consider using PID control for temperature. In that case, the above discussion can be displayed in the graph below:

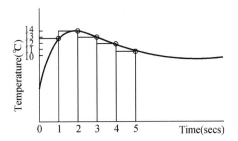

Figure 2-98 PID for Temperature[23]

In the above graph, the graph represents the temperature, points on the graph are the point that temperature was sampled, and blue areas are integral of the temperature signal.

PID Application Example

An essential practical example in the robotics lab is a line follower robot. We aim to keep the robot smoothly moving alongside a black line with appropriate space. For this example, we consider an IR line sensor with five sensors. IR sensors consist of an IR transmitter and receiver. Light colour surfaces reflect IR emissions, while black limits or cannot reflect it at all.

Figure 2-99 Five Channel IR Tracker

To achieve a better result with PID, the black tape must cover an area of a minimum of one sensor and a maximum of two sensors. Each IR sensor may return '1' or '0' on the black surface based on the type of sensor. Our selected device returns '1' on the black line and '0' on the white (or light reflection colours). While the car is moving alongside the black line, the following conditions may occur:

$$0\ 0\ 1\ 0\ 0$$
$$0\ 0\ 0\ 0\ 1$$
$$0\ 0\ 0\ 1\ 1$$
$$0\ 0\ 0\ 1\ 0$$
$$0\ 0\ 1\ 1\ 0$$
$$0\ 1\ 1\ 0\ 0$$
$$0\ 1\ 0\ 0\ 0$$
$$1\ 1\ 0\ 0\ 0$$
$$1\ 0\ 0\ 0\ 0$$

The set point in this example is the first condition in this list. The error variable related with the sensor status will be:

0 0 1 0 0 = = > Error = 0
0 0 0 0 1 = = > Error = 4
0 0 0 1 1 = = > Error = 3
0 0 0 1 0 = = > Error = 2
0 0 1 1 0 = = > Error = 1
0 1 1 0 0 = = > Error = -1
0 1 0 0 0 = = > Error = -2
1 1 0 0 0 = = > Error = -3
1 0 0 0 0 = = > Error = -4

The process of reducing errors and returning the robot's centre IR sensor on the black line (error = 0) is demonstrated in the image below:

Chapter 2　Introduction to Embedded Robotics

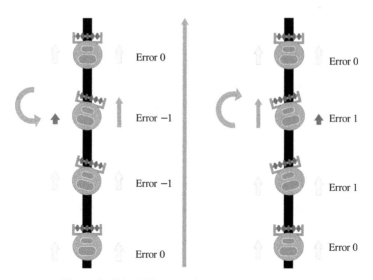

Figure 2-100　Using PID in Line Follower Robot

The PID process would assist us in moving the robot to the correct position by using a control loop feedback mechanism to control process variables.

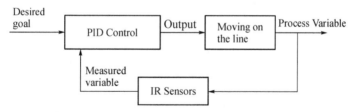

Figure 2-101　PID Process Diagram

The system calculates the 'error' or 'deviation' of the physical quantity from the set point by measuring the current value of that physical quantity using a sensor(s). To get back to the set point, this 'error' should be minimized and ideally equal to zero. Also, this process should happen as quickly as possible. Ideally, there should be zero lag in the system's response to the change in its set point.

Implementing PID

ⅰ) Error Term (e)

This is equal to the difference between the set point and the current value of the quantity being controlled.

$$\text{Error} = \text{setpoint} - \text{current value}$$

(In our case is the error variable obtained from the position of robot over the line)

ii) Proportional Term (P)

This term is proportional to the error.

$$P = \text{error}$$

This value is responsible for the magnitude of change required in the physical quantity to achieve the set point. The proportion term determines the control loop rise time or how quickly it will reach the set point.

iii) Integral Term (I)

This term is the sum of all the previous error values.

$$I = I + \text{error}$$

This value is to achieve the speed of response to the change from the set point. The integral value is to eliminate the steady state error required by the proportional term. Usually, small robots do not need the integral term because in more miniature robots, we are not concerned about the steady-state error, which can complicate the "loop tuning".

iv) Differential or Derivative Term (D)

This term is the difference between the instantaneous error from the set point and the previous instant.

$$D = \text{error} - \text{previous Error}$$

This value is responsible for slowing down the rate of change of the physical quantity when it comes close to the set point. The derivative term reduces the overshoot or how much the system should "overcorrect".

Implementing PID equation:

$$PID \text{ value} = (K_p \times P) + (K_i \times I) + (K_d \times D)$$

In this example, we can consider three coefficients:

K_p is the constant used to vary the magnitude of the change required to achieve the set point.

K_i is the constant used to vary the rate at which the change should be brought in the physical quantity to achieve the set point.

K_d is the constant used to vary the stability of the system.

If the robot responses reasonable with a $K_p = 50$, then set $K_p = 25$ and $K_d = 25$ to start. Increase the K_d (derivative) gain to decrease the overshoot, and decrease it if the robot becomes unstable.

- If the robot reacts too slowly, increase the value.
- If the robot seems to react quickly and becomes unstable, decrease the value.

Another component of the loop is the actual *Sample/Loop Rate*. Speeding this parameter up or slowing this down can make a significant difference in the robot's performance.

PWM

A PWM (pulse width modulation) signal is a digital square wave with constant frequency but a different duty cycle, which can be varied between 0% and 100%.

Pulse width modulation (PWM) is used in various applications, including sophisticated control circuitry. Using PWM, a range of applications, including control of dimming RGB LEDs, speed of DC motors, or control of the direction of a servo motor, can be achieved. Those applications benefit from the basic principle of PWM, which is by deciding how much time the signal is to be HIGH. By changing the proportion of time, the signal is HIGH to LOW in a consistent time, the signal can be HIGH (usually 5 volts) at any interval.

Duty Cycle

The term "on time" refers to when the signal strength is optimal. The term "duty cycle" refers to the percentage of "on time" in a certain period. The duty cycle is the ratio of available time to the total time the device is on. The % duty cycle is a measure of how much of the time a digital signal is active within a given period. The frequency is inverse of its period.

The perfect square wave has a duty cycle of 50%, a digital signal that's on for half the time and off for the other half. Put another way, if the duty cycle is greater than 50%, the digital signal is in the high state more often than the low state, and vice versa if it is less than 50%.

At a voltage of 5 volts, and a duty cycle of 100%, it would be the same as if the voltage had been set to 5 volts. On the other hand, having a duty cycle of 0% is the same as having the signal grounded.

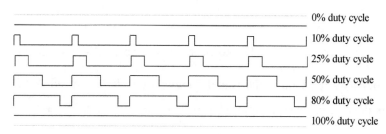

Figure 2-102　Comparing Different Duty Cycles in PWM

To dim LEDs properly, we need to make sure the frequency of the square wave is high enough. To the naked eye, a 20% duty cycle wave at 1 Hz looks like it's blinking on and off, while at 100 Hz or above, it simply appears dimming than when it's at full power. For an appropriate dimming effect, the period can't be too long.

PWM Examples

PWM has several uses:

- Dimming an LED;
- Providing an analog output; if the digital output is filtered;
- Providing an analog voltage between 0% and 100%;
- Generating audio signals;
- Providing variable speed control for motors;
- Generating a modulated signal, for example to drive an infrared LED for a remote control.

Simple Pulse Width Modulation with Arduino

Using Arduino's programming language is easy to make PWM signals as we need. We can generate a PWM signal using analogWrite(pin, duty cycle) at the provided PWM pins (Arduino UNO: 3, 5, 6, 9, 10, 11). It must be mentioned that despite the name 'analogue', the result is a digital signal. Here duty cycle can be 0 to 255. Although analogWrite() provides an easy way to generate PWM signals, it doesn't give us any control over frequency. The Fading example demonstrates how to use the analogue output (PWM) to control an LED illumination. The example program is available from Arduino's 'Examples/Analog' list from the 'File' menu.

Complete Arduino program

```
/*
Fade
This example shows how to fade an LED on pin 9
using the analogWrite() function.
This example code is in the public domain.
*/

int led = 9;           // the pin that the LED is attached to
int brightness = 0;    // how bright the LED is
int fadeAmount = 5;    // how many points to fade the LED by

// the setup routine runs once when you press reset:
void setup() {
  // declare pin 9 to be an output:
  pinMode(led, OUTPUT);
}


// the loop routine runs over and over again forever:
void loop() {
```

```
// set the brightness of pin 9:
analogWrite(led, brightness);

// change the brightness for next time through the loop:
brightness = brightness + fadeAmount;

// reverse the direction of the fading at the ends of the fade:
if (brightness = = 0 || brightness = = 255) {
  fadeAmount = - fadeAmount;
}

// wait for 30 milliseconds to see the dimming effect

delay(30);

}
```

Controlling Servo Motors

The frequency (period) is specific to controlling a specific servo. For instance, a servo motor may require a PWM signal to be updated every 20 ms with a pulse between 1 ms and 4 ms or between a 5% and 20% duty cycle on a 50 Hz. With a 1.5 ms pulse, the servo will move to the natural 90-degree position; when we feed with a 1 ms pulse, the servo will be at the 0-degree position; with a 4 ms pulse, the servo will move to 180 degrees. The full range of motion can be obtained by updating the servo with a value between the range. In a mechanical robot arm, servos have a shaft that turns to a specific position within its range (usually 0-180 degrees).

Chapter 2 Introduction to Embedded Robotics

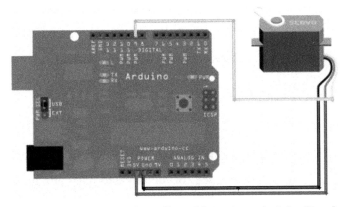

Figure 2-103   Connecting a Servo Motor to an Arduino Board

/ * Sweep

by BARRAGAN < http: //barraganstudio. com >

This example code is in the public domain.

modified 8 Nov 2013

by Scott Fitzgerald

http: //www. arduino. cc/en/Tutorial/Sweep

* /

#include < Servo. h >

Servo myservo; // create servo object to control a servo

// twelve servo objects can be created on most boards

int pos = 0; // variable to store the servo position

void setup( ) {

  myservo. attach(9) ; // attaches the servo on pin 9 to the servo object

}

void loop( ) {

  for (pos = 0; pos < = 180; pos + = 1) { // goes from 0 degrees to 180 degrees

```
                                        // in steps of 1 degree
    myservo.write(pos);      // tell servo to go to position in variable 'pos'
    delay(15);               // waits 15ms for the servo to reach the position
  }
  for (pos = 180; pos >= 0; pos -= 1) { // goes from 180 degrees to 0 degrees
    myservo.write(pos);      // tell servo to go to position in variable 'pos'
    delay(15);               // waits 15ms for the servo to reach the position

  }

}
```

Manipulators-I

Manipulators have been focused on academia and small industries during recent decades due to improved technology and lower prices. Until about ten years ago, only big industries had the opportunity to use robotics arms in the production line, packaging, or sorting. However, recently, there are even versions for home users because of improvements to control methods and lower prices. At the same time, there is increasing research in robotic arm-related topics. Without a doubt, after 3-D printers, the robotics arm will be the next generation widely used in all industries and education.

Its precise positioning with repeatability factors makes the robotics arm reliable, which lets them be used in production lines, 3-D printers, assembly lines, robotics, safety, and more. Forward and Inverse Kinematics are the essential starting points at the heart of the robotics arm's movement and positioning.

Chapter 2 Introduction to Embedded Robotics

Talking about Kinematics, we must refer to mathematics, physics and classic mechanical engineering. There are two mainly used Kinematics in the robotic field: forward Kinematics and inverse Kinematics.

The frequently used forward Kinematics is to calculate the position of the end effector when the degree value of each joint is known, and inverse Kinematics is to compute the degree of each joint when we know the position of the end effector.

$$\cos\theta = x/l$$
$$x = \cos\theta \times l = \cos 35° \times 8 \text{ cm}$$
$$\sin\theta = y/l$$
$$y = \sin\theta \times l = \sin 35° \times 8 \text{ cm}$$

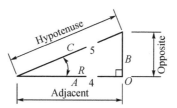

Table 2-11 Forward Kinematics Calculation

α	0°	30°	45°	60°	90°
$\sin\alpha$	$\frac{1}{2}\sqrt{0}$	$\frac{1}{2}\sqrt{1}$	$\frac{1}{2}\sqrt{2}$	$\frac{1}{2}\sqrt{3}$	$\frac{1}{2}\sqrt{4}$
$\cos\alpha$	$\frac{1}{2}\sqrt{4}$	$\frac{1}{2}\sqrt{3}$	$\frac{1}{2}\sqrt{2}$	$\frac{1}{2}\sqrt{1}$	$\frac{1}{2}\sqrt{0}$
$\tan\alpha$	$\sqrt{\frac{0}{4}}$	$\sqrt{\frac{1}{3}}$	$\sqrt{\frac{2}{2}}$	$\sqrt{\frac{3}{1}}$	$\sqrt{\frac{4}{0}}$

Example code in python:

```
#!/usr/bin/env python3
"
Basic robotic forward kinematics solver
 Huaiyin Institute of Technology
 Robotics Research Labs


 Amir
```

```
import math
def InversKin1dof(_link, _degree):
    try:
        x = 0
        y = 0
        x = math.cos(math.radians(_degree)) * _link
        y = math.sin(math.radians(_degree)) * _link
        return x, y
    except:
        raise
x, y = InversKin1dof(8, 35)
print("x = " + str(x) + " y = " + str(y))
```

Result:

By running the above program for x = 8, and y = 35:

x = 6.553216354311934 y = 4.588611490808368

Then:

Position of end effector = p(6.55 cm, 4.59 cm)

Forward Kinematics of a 3dof robot arm to obtain end factor position:

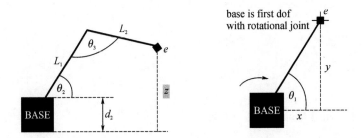

Figure 2-104　Forward Kinematics of a 3dof Robot Arm

Where: d_2 is the height of the second joint towards the ground, and z is the height of the end effector from the ground.

Chapter 2 Introduction to Embedded Robotics

L_1 is the length of link 1, and L_2 is the length of link 2.

θ_1 is d_1 joint value, d_2 joint's value is θ_2, and θ_3 is the d_3 joint value.

We consider our cartesian coordinate from the top view of our robotic arm.

We will calculate the end effector (E) position at 3-dimensional spaces (x, y, z).

Our known variables: °

$L_1 = L_2 = 10$ cm
$d_2 = 7$ cm
$\theta_1 = 60°$
$\theta_2 = 30°$
$\theta_3 = 140°$

We mark some more degrees and lengths:

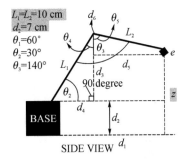

Figure 2-105 Side View of Forward Kinematics of a 3dof Robot Arm

Steps:

since $z = d_2 + d_3 - d_6$, first step is finding d_3

Step 1. finding d_3

$\sin\theta_2 = d_3/L_1$

$\sin 30° = d_3/10$

$d_3 = \sin 30° \times 10 = 4.9$ cm

Next step is finding d_2 and d_6 length. In order to get d_2 and d_6 length, we need to get more information.

Step 2. finding θ_4

$180° = \theta_2 + 90° + \theta_4$

$\theta_4 = 180° - (30° + 90°) = 60°$

Step 3. find θ_5

$\theta_3 = \theta_4 + \theta_5$

$140° = 60° + \theta_5$, hence $\theta_5 = 80°$

Step 4. finding d_6 and z

Since we have θ_5, finding d_6:

$\cos\theta_5 = d_6/L_2$

$\cos 80° = d_6/10$

$d_6 = \cos 80° \times 10 = 1.7$ cm

So, we have below information:

$z = 7$ cm $+ 4.9$ cm $- 1.7$ cm $= 10.2$ cm

Next, we need to find x and y.

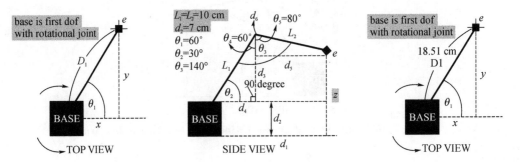

Figure 2-106 Side and Top View of 3dof Robot Arm

x and y can be obtained if we got the hypotenuse length (d_1).

The side view that $d_1 = d_4 + d_5$

Step 5. finding d_4

$\cos\theta_2 = d_4/L_1$

$d_4 = \cos 30° \times 10 = 8.66$ cm

step 6. finding d_5 and d_1

since we have $\theta_5 = 80°$

$\sin 80° = d_5/10$, so $d_5 = \sin 80° \times 10 = 9.85$ cm

$d_1 = d_4 + d_5 = 8.66$ cm $+ 9.85$ cm $= 18.51$ cm

On the top view, we have found x and y.

$\cos\theta_1 = x/d_1$

$\cos 60° = x/18.51$

$x = \cos 60° \times 18.51 = 9.25$ cm

$\sin 60° = y/18.51$, $y = \sin 60° \times 18.51 = 16.03$ cm

Finally we find that $p(x, y, z) = p(9.25, 16.03, 10.2)$.

Advance Use of Microcontrollers

Dealing with 'Delay'

For robotics hobbyists, before mastering one technique, they often jump to another exciting subject area. Ease of access to the public shared resources widely available through the Internet makes this jump even easier without needing a deep understanding. However, with academic demands, students face more challenges and require mastering each concept, theoretical and practical skills. For instance, Blink, one of the simplest examples of Arduino programs, can be discussed in more detail to understand more advanced programming of Arduino microcontrollers.

Arduino delay command:

Blinking LEDs is an exciting project for learning about digitalWrite and 'delay' functions. However, moving forward and having more LEDs with different 'delays' simultaneously is challenging using these functions. Because Arduino runs on a simple microprocessor with no operating system, so it is designed to run a single task simultaneously. Unlike Raspberry Pi or another microcontroller, Arduino cannot run multiple programs. For instance, the blink program only can blink an LED at a time:

```
1.  /*
2.  Blink
3.  Turns on an LED on for one second, then off for one
4.  second, repeatedly.
5.  This example code is in the public domain.
6.  */
7.  // Pin 13 has an LED connected on most Arduino boards.
8.  // give it a name:
9.  int led = 13;
10. // the setup routine runs once when you press reset:
11. void setup() {
12. // initialize the digital pin as an output.
13. pinMode(led, OUTPUT);
14. }
15. // the loop routine runs over and over again forever:
16. void loop() {
17. digitalWrite(led, HIGH); // turn the LED on (HIGH is the
18. voltage level)
19. delay(1000); // wait for a second
20. digitalWrite(led, LOW); // turn the LED off by making the
21. voltage LOW
22. delay(1000); // wait for a second
23. }
24.
```

In the above program, most of the time is spent on delay function or, in other words sleeping. It is the same with the 'sweep' program to continuously move a servo motor from 0 degrees to 180 degrees.

The 'delay' function pauses the microcontroller from any tasks for the designated period. For instance, using delay(1000), pause your program for a second. Therefore, to deal with multiple tasks, there must be another way of timekeeping.

Fortunately, there is another way of keeping time without delaying the processor from other tasks. Using the "millis()" (milliseconds = 1/1000 second) function, we can control flashing an LED. "millis()" returns the number of seconds since the Arduino started running the program. If we do not stop the program, this number returns to zero after about 50 days. The data type for holding the "millis()" value is (unsigned long), which stores 32 bits (4 bytes). Unlike a standard long variable, unsigned long will not store negative numbers. The range of stored values is from 0 to 4,294,967,295 ($2^{32}-1$).

Chapter 2   Introduction to Embedded Robotics

A sample program to demonstrate millis( ) function:

```
1. unsigned long time;
2. void setup(){
3.   Serial.begin(9600);
4. }
5. void loop(){
6.   time = millis();
7.   Serial.print("Time: ");
8.   Serial.println(time);
9. }
```

To replace the delay ( ) function with millis( ), we need to record the current millis( ) value and start to count until the desired time arrives.

```
1. …
2. long previousMillis = 0;
3. unsigned long currentMillis = millis();
4. if(currentMillis - previousMillis > interval) {
5. previousMillis = currentMillis;
6. // Do something
7. ….
8. }
9. …
```

The complete Blink program using millis can be written as:

```
1.  /* Blink without Delay
2.  Turns on and off a light emitting diode(LED) connected to a digital
3.  pin, without using the delay() function.
4.  created 2005
5.  by David A. Mellis
6.  modified 8 Feb 2010
7.  by Paul Stoffregen
8.  This example code is in the public domain.
9.  http://www.arduino.cc/en/Tutorial/BlinkWithoutDelay
10. */
11. // constants won't change. Used here to
12. // set pin numbers:
13. const int ledPin = 13;
14. // the number of the LED pin
15. // Variables will change:
16. int ledState = LOW;
17. long previousMillis = 0;
18. // ledState used to set the LED
19. // will store last time LED was updated
20. // the follow variables is a long because the time, measured in
21. miliseconds,
22. // will quickly become a bigger number than can be stored in an int.
23. long interval = 1000;
24. // interval at which to blink (milliseconds)
25. void setup() {
26. // set the digital pin as output:
27. pinMode(ledPin, OUTPUT);
28. }
29. void loop()
30. {
31. // here is where you'd put code that needs to be running all the
32. time.
33. // check to see if it's time to blink the LED; that is, if the
34. // difference between the current time and last time you blinked
35. // the LED is bigger than the interval at which you want to
36. // blink the LED.
37. unsigned long currentMillis = millis();
38. if(currentMillis - previousMillis > interval) {
39. // save the last time you blinked the LED
40. previousMillis = currentMillis;
41. // if the LED is off turn it on and vice-versa:
42. if (ledState == LOW)
43. ledState = HIGH;
44. else
45. ledState = LOW;
46. // set the LED with the ledState of the variable:
47. digitalWrite(ledPin, ledState);
48. }
49. }
```

Here, we come to a point and learn another vital subject in programming intelligent, practical robots. That is defining the 'state' of things. What creates intelligence is memory.

Chapter 2 Introduction to Embedded Robotics

By investigating the natural world, we can see animals having relatively advanced memory and can enjoy a higher level of intelligent decision-making.

For a robot at each stage to make a decision, it must be aware of its current and previous state.

The above code, as much as it is more complicated with the code, using the delay( ) function brings us two important things. First, during the counting times for the LED state, the processor can continue with other possible tasks; second, at any moment, it can tell us the state of the LED (is it ON or OFF?). That is the state a robot needs to be aware of its condition[24].

Another example of the Blink code by Adafruit: Here, we use LED off-time and on-time at two different times. It is similar to using two different 'delay( )' functions. Although it seems more complicated than using the delay( ) function, it is a straightforward concept; to check the current LED state and the time passed to change the LED state.

```
1. long onTime = 250; // milliseconds of on-time
2. long offTime = 750; // milliseconds of off-time
3. …
4. void loop()
5. {
6. // check to see if it's time to change the state of the LED
7. unsigned long currentMillis = millis();
8. if((ledState == HIGH) && (currentMillis - previousMillis >= onTime))
9. {
10. ledState = LOW; // Turn it off
11. previousMillis = currentMillis; // Remember the time
12. digitalWrite(ledPin, ledState); // Update the actual LED
13. }
14. else if ((ledState == LOW) && (currentMillis - previousMillis >=
15. offTime))
16. {
17. ledState = HIGH; // turn it on
18. previousMillis = currentMillis; // Remember the time
19. digitalWrite(ledPin, ledState); // Update the actual LED
20. }
21. }
22. …
```

Multitasking and OO Programming

Another situation we need to consider in more advanced programming is multi-tasking. A microcontroller like the one used with the Arduino development board does not support multi-tasking directly. Therefore, we can run only one task at a time. We can run a servo motor or check the ultrasonic range finder. With more extensive programs and more devices connected to an Arduino development board, we may consider moving to object-oriented programming.

Arduino development board does not support OO by default; however, as the programming environment is a version of C++, then it can support Object Oriented Programming too.

Instead of writing functions and subroutines, we write classes for each behaviour, sensor, or controlled device.

Object Oriented Programming with Microcontrollers

A reader unfamiliar with object-oriented programming should read this section.

The first concept in object-oriented programming is 'object'. Objects, as their name in English refer to, can be anything and everything. Everything we see or imagine can be considered an object. Objects have their properties, such as physical object's properties are size, colour, material, weight and more.

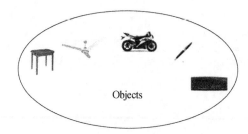

Figure 2-107 Objects

Chapter 2   Introduction to Embedded Robotics

In the natural world, we may use 'class' to refer to species, a group of similar kinds, such as mammals and birds. Of course, there are different mammals in that class. However, they all share the same characteristic as the class identify it, which is being a 'mammal'. Members of each class, birds or mammals, have the same characteristics to be part of their group, but at the same time, they may differ in some other behaviours. For instance, all mammals give birth, as objects' natures may be very different from each other, sometimes we group them with a specific characteristic. This grouping helps us be able to deal with them more appropriately.

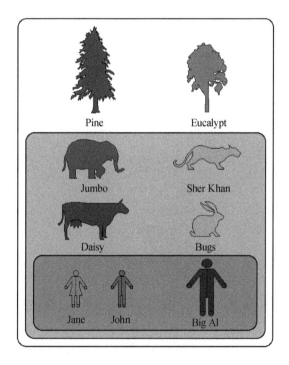

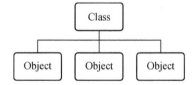

Figure 2-108   Classes and Objects

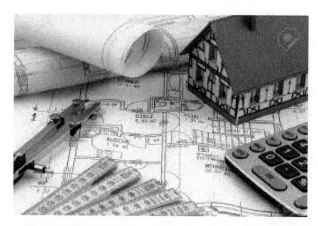

Figure 2-109 A Blueprint of a House Is Not a House

A car manufacturer decides to make a new car. First, they ask designers to design the car with the desired specification on paper or computer. This design does not include specific information regarding the car's body colour, nor are various functions available with different models of the same car. Instead, this design is a blueprint of the idea.

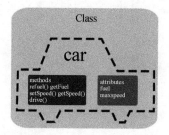

Later, after acceptance of the blueprint, the manufacturer sends the design to the production unit to make many cars from this blueprint; each may vary with colour, number of doors, or engine size.

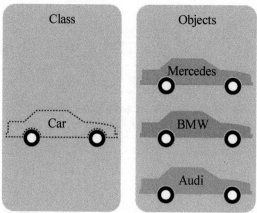

Characteristics (properties) and behaviours (methods) that each object has made them individual instances of the same class.

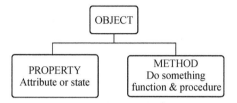

In object-oriented view, the car's blueprint is the car class. Car class is not a car; it is a design. However, every instance of an actual car being made in a production unit is an object of that car.

Writing object-oriented programs starts with writing a class file. In C++ class is defined by:

```
class MyClass
{
//class variables
// Here you write class variables


public:              // This is constructor. Telling the class how to make each
MyClass(int a, long b, …)
{
// Constructor is similar to Arduino setup() function
// usually what is in setup, goes in here
}

void Update()
{
// This part can be consider similar to loop() in Arduino programming
}
```

After writing a class file, then creating instances (objects) of that class as easy as:

MyClass. myObject (a, b, …)

MyClass. myObject2 (a, b, …)

Here are the blinking LEDs in different timing sketch by Adafruit:

The Arduino language is a variant of C++ which supports object-oriented programming. Using the OOP features of the language, we can gather all of the state variables and functionality for a blinking LED into a C++ class. This is relatively easy to do, and we have already written all the code for it and need to re-package it as a class. Defining a class: We start by declaring a "Flasher" class; then, we add all the variables from Flash Without Delay. Since they are part of the class, they are known as member variables.

```
1.  class Flasher {
2.  // Class Member Variables
3.  // These are initialized at startup
4.  int ledPin;
5.  // the number of the LED pin
6.  long OnTime; // milliseconds of on-time
7.  long OffTime; // milliseconds of off-time
8.  // These maintain the current state
9.  int ledState;
10. // ledState used to set the LED
11. unsigned long previousMillis; // will store last time LED was updated
12. }
```

Writing a constructor is relatively easy as it is an Arduino setup. All the input/output settings and initial states will occupy construction. Same as in the Arduino setup ( ) function, the constructor only will be executed during each object construction. Construction also initialises each object by providing variables pin, on, and off. Then it is possible to create any object connected to an Arduino pin, specifying 'on' time and 'off' time flashing periods.

```
14. public:
15. Flasher(int pin, long on, long off)
16. {
17. ledPin = pin;
18. pinMode(ledPin, OUTPUT);
19. OnTime = on;
20. OffTime = off;
21. ledState = LOW;
22. previousMillis = 0;
23. }
```

Then the final task is to write the primary expected behaviour or the task we want an object to perform.

Chapter 2   Introduction to Embedded Robotics

```
24. void Update()
25. {
26. // check to see if it's time to change the state of the LED
27. unsigned long currentMillis = millis();
28. if((ledState == HIGH) && (currentMillis - previousMillis >= OnTime))
29. {
30. ledState = LOW; // Turn it off
31. previousMillis = currentMillis; // Remember the time
32. digitalWrite(ledPin, ledState); // Update the actual LED
33. }
34. else if ((ledState == LOW) && (currentMillis - previousMillis >= OffTime))
35. {
36. ledState = HIGH; // turn it on
37. previousMillis = currentMillis; // Remember the time
38. digitalWrite(ledPin, ledState); // Update the actual LED
39. }
40. }
41. };
```

The final task is to create as many objects as we want and can connect to the Arduino's digital input pins. The result is a small and elegant program.

```
1. Flasher led1(12, 100, 400);
2. Flasher led2(13, 350, 350);
3. void setup()
4. {
5. }
6. void loop()
7. {
8. led1.Update();
9. led2.Update();
10. }
```

User Interfaces

User interfaces for microcontrollers have more limitations than SBC (Single Board Computer) development boards. SBC can usually employ the same user interfaces as PCs, but for microcontrollers, this option is limited by size, power, connectivity, and software limitation. On the other hand, embedded robotics projects require a minimum of user interfacing. This section provides some of such interfaces and their connection diagram and software code to complete our task.

I2C LCD 2/4 Line

Integrating an LCD facilitates the project's interactivity, allowing the user to read some output parameters directly. These values can be either simple text or numerical values read by the sensors, such as temperature or pressure or even the number of cycles the Arduino performs.

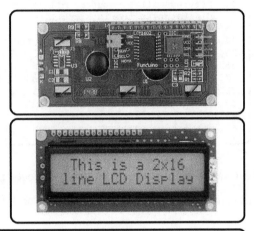

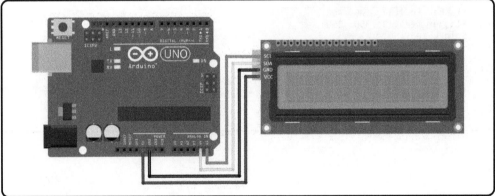

Chapter 2 Introduction to Embedded Robotics

```
#include <LiquidCrystal_I2C.h>

char row1[]="I'm Arduino";
char row2[]="Hello world!";
int t = 500;
// Library initialization with 16x2 size settings
LiquidCrystal_I2C lcd(0x27,16,2); // set the LCD address to 0x27

void setup(){
   lcd.init();
   lcd.backlight();
}

void loop()
{
    lcd.clear();
    lcd.setCursor(15,0);
    for (int positionCounter1 = 0; positionCounter1 < 26; positionCounter1++){
        lcd.scrollDisplayLeft();
        lcd.print(row1[positionCounter1]);
        delay(t);
    }
    lcd.setCursor(15,1); // set the cursor to column 15, line 1
    for (int positionCounter = 0; positionCounter < 26; positionCounter++){
        lcd.scrollDisplayLeft();
        lcd.print(row2[positionCounter]);
        delay(t);
    }
}
```

Exercises and Projects

Changes in the field of robotics are speedy. Therefore, to provide more UpToDate, more up-to-date problem-solving techniques and innovative projects, this part will be available from the author's website at www. roboticacamp. com/robotics_and_ai.

Exercises and Projects

Chapter 3

Robotics Design and Control

Robotics control is an essential system in robot movement which involves the mechanical aspects and program system. Depending on the type of robot, traditional control systems can provide manual, semi-autonomous, or fully autonomous control.

During the last decade, more and more robotics control systems have even moved toward intelligent control. This is valid both for mobile robots and even robotics arms. The difference between automation and intelligence is quite simple. Automation using various methodologies can provide a program for predefined methods and ways to perform a specific task. However, intelligent design using machine learning expects the robot to make an independent decision based on unexpected changes to complete the same task.

Different Robots

One of the main categories of robots is indoor and outdoor robots. Here our focus is on indoor robots.

Indoor mobile robots are becoming more and more popular in education and industry. By making an indoor definition, we only mean a robot is only designed to operate indoors. One of the most affordable indoor robots for education and yet provide most of the robotics education needs is Turtlebots.

Indoor robots usually have a 3D camera or a lidar for SLAM (Simultaneous Localisation and Mapping). They can be wheeled robots or humanoids (walking on two legs).

Turtlebot was one of the successors of educational indoor robots. Since the first Turtlebot-1 used a Rambo hoover robot, the main goal was to provide a programmable robot for education. Turtlebot-2 uses a Kabuki robot base with more accurate odometry.

Turtlebot-3 uses its robot base with servo motors and an OpenCR controller from Robotis company.

Another kind of robot becoming more and more popular is AGV (Automated Guided Vehicle).

Tools for Intelligent Robotics Design

Image Processing

Image processing is one of the essential parts of machine vision. Image analysis always requires image processing as the captured images do not always contain the visible data we need. Each project may need different data types to extract from an image. This section reviews the most common and valuable image processing method in robotics.

OpenCV

The "computer vision" library contains many different computer vision functions and machine learning functions. The library has more than 2,500 optimized algorithms, including a comprehensive set of classic and state-of-the-art computer vision and machine learning algorithms. It was created by Intel and was initially released in 2000. OpenCV has many different algorithms related to computer vision that can perform a variety of tasks, including facial detection and recognition, object identification, monitoring moving objects, tracking camera movements, tracking eye movements, extracting 3D models of objects, creating an augmented reality overlay with a scenery, recognizing similar images in an image database, and more.

OpenCV has many applications including:

- 2D and 3D feature toolkits
- Egomotion estimation
- Facial recognition system

Chapter 3 Robotics Design and Control

- Gesture recognition
- Human-computer interaction (HCI)
- Mobile robotics
- Motion understanding
- Object identification
- Segmentation and recognition
- Stereopsis stereo vision: depth perception from 2 cameras
- Structure from motion (SFM)
- Motion tracking
- Augmented reality

OpenCV is one of the essential machine vision libraries used in robotics. It has more than 47,000 people in the user community and an estimated number of downloads exceeding 18 million or 180 lakh, which makes it more critical in the world community[25].

Steps in Image Processing

Here we start with basic image processing processes using OpenCV. Here we will use python with OpenCV. However, OpenCV online tutorial has all codes both for Python and C++.

Open, Display, Save

After installing OpenCV, the first step is to check if we have access to the OpenCV library in Python by starting Python and typing the following line. If there is no error, then we have access to Python.

import cv2 as cv

Then the first step in image processing is to open an image, display and save it.

```
1. import cv2 as cv
2. import sys
3. img = cv.imread(cv.samples.findFile("test2.jpg"))
4. if img is None:
5.     sys.exit("Could not read the image.")
6. cv.imshow("Display window", img)
7. k = cv.waitKey(0)
8. if k == ord("s"):
9.     cv.imwrite("test2.png", img)
```

Line 1, and 2 import openCV and sys libraries.

Line 3, loading image from 'test2.jpg'

Line 4, and 5, for errors with loading the image

Line 6, to display the image img

Line 7, waiting for a key-press

Line 8, and 9, if a key is pressed, the image will be saved in png format

To load a more complicated video. Here the code is already commented on and does not need further explanation.

```
1.  import numpy as np
2.  import cv2 as cv
3.  cap = cv.VideoCapture(0)
4.  if not cap.isOpened():
5.      print("Cannot open camera")
6.      exit()
7.  while True:
8.      # Capture frame-by-frame
9.      ret, frame = cap.read()
10.     # if frame is read correctly ret is True
11.     if not ret:
12.         print("Can't receive frame (stream end?). Exiting ...")
13.         break
14.     # Our operations on the frame come here
15.     gray = cv.cvtColor(frame, cv.COLOR_BGR2GRAY)
16.     # Display the resulting frame
17.     cv.imshow('frame', gray)
18.     if cv.waitKey(1) == ord('q'):
19.         break
20. # When everything done, release the capture
21. cap.release()
22. cv.destroyAllWindows()
```

Playing video from a file:

```
1. import numpy as np
2. import cv2 as cv
3. cap = cv.VideoCapture('vtest.avi')
4. while cap.isOpened():
5.     ret, frame = cap.read()
6.     # if frame is read correctly ret is True
7.     if not ret:
8.         print("Can't receive frame (stream end?). Exiting ...")
9.         break
10.     gray = cv.cvtColor(frame, cv.COLOR_BGR2GRAY)
11.     cv.imshow('frame', gray)
12.     if cv.waitKey(1) == ord('q'):
13.         break
14. cap.release()
15. cv.destroyAllWindows()
```

```
1. import numpy as np
2. import cv2 as cv
3. cap = cv.VideoCapture(0)
4. # Define the codec and create VideoWriter object
5. fourcc = cv.VideoWriter_fourcc(*'XVID')
6. out = cv.VideoWriter('output.avi', fourcc, 20.0, (640, 480))
7. while cap.isOpened():
8.     ret, frame = cap.read()
9.     if not ret:
10.         print("Can't receive frame (stream end?). Exiting ...")
11.         break
12.     frame = cv.flip(frame, 0)
13.     # write the flipped frame
14.     out.write(frame)
15.     cv.imshow('frame', frame)
16.     if cv.waitKey(1) == ord('q'):
17.         break
18. # Release everything if job is finished
19. cap.release()
20. out.release()
21. cv.destroyAllWindows()
```

The below code captures from a camera, flips (line 12) every frame in the vertical direction, and saves the video.

The captured camera (line 3), will be saved as 'output. avi' with 20 fps (frame per second) and 640 × 480 pixels in size.

To draw a line:

```
1. import numpy as np
2. import cv2 as cv
3. # Create a black image
4. img = np.zeros((512,512,3), np.uint8)
5. # Draw a diagonal blue line with thickness of 5 px
6. cv.line(img,(0,0),(511,511),(255,0,0),5)
```

Drawing other objects, adding text:

```
1.  # To draw a rectangle
2.  cv.rectangle(img,(384,0),(510,128),(0,255,0),3)
3.
4.  # To draw a circle
5.  cv.circle(img,(447,63), 63, (0,0,255), -1)
6.
7.  # To draw a ellipse
8.  cv.ellipse(img,(256,256),(100,50),0,0,180,255,-1)
9.
10. # To draw a polygon
11. pts = np.array([[10,5],[20,30],[70,20],[50,10]], np.int32)
12. pts = pts.reshape((-1,1,2))
13. cv.polylines(img,[pts],True,(0,255,255))
14.
15. # Adding text
16. font = cv.FONT_HERSHEY_SIMPLEX
17. cv.putText(img,'OpenCV',(10,500), font, 4,(255,255,255),2,cv.LINE_AA)
```

Further OpenCV functions and tools can be accessed from https://docs.opencv.org/4.x/d6/d00/tutorial_py_root.html. It is a fact that OpenCV.org has already made one of the best documentation, and there is no need to repeat the same.

Chapter 3 Robotics Design and Control

Project: GAPCAL (2019–2021)

GAPCAL—Contactless gap and flush calculation

GAPCAL is a contactless gap and flush calculation device developed in our robotics AI labs. The working principle is based on laser triangulation. The device uses a Raspberry Pi, and Arduino Nano, with Pylon CMOS cameras.

All packages are written in Python and C++. It uses OpenCV image processing. Using an innovative IMCM method developed by Amir Ali Mokhtarzadeh, the accuracy of gap and flush calculation reached 0.02 mm.

ROS (Robot Operating System)

ROS is an acronym for 'Robot Operating System'. ROS is a set of software libraries and tools that helps build robot applications. ROS is used by students, researchers, and industry, and ROS is open source.

Many people new to ROS may need to learn more about that particular hardware to learn how ROS lets developers write the program for their hardware. ROS is an extensive collection of libraries and tools that connect our hardware and orchestrate with another part of our robot in just a few steps. That may be why after using ROS, we may not be able to do without it.

ROS is not an alternative operating system to Windows or Linux. It goes on top of our operating system (most commonly Ubuntu Linux).

ROS is becoming the standard in robotics programming, at least in the service robot sector. Initially, ROS started at universities but quickly spread into the business world. Every day, more and more companies and start-ups are basing their businesses on ROS. Before ROS, every robot had to be programmed using the manufacturer's API. If we change our robots, we must start the entire software again, apart from learning the new API[26].

Do we need ROS? Maybe not. However, with ROS, we are joined and writing all software tools, simulations, and communication between different packages and libraries. There is no need to mention ROS community support.

ROS is supported best for Ubuntu and Debian and experimental versions for Gentoo Linux, OS X (Homebrew), and Open Embedded.

ROS versions from the beginning (2010) are:

- ROS Turtle
- Box Turtle
- C Turtle
- Diamondback

Chapter 3 Robotics Design and Control

- Electric Emys
- Fuerte Turtle
- Groovy Galapagos
- Hydro Medusa
- Indigo Igloo
- Jade Turtle
- Kinetic Kame
- Lunar Loggerhead
- Melodic Morenia
- Noetic Ninjemys

Also, ROS2 versions started from 'alpha1' release in 2015. At the time of writing this book, the final version is 'Humble Hawksbill'.

After installing ROS, you may need to check environment by:

$$\text{printenv | grep ROS}$$

Also, to make sure your operating system has access to ROS installation. You must run the command below:

$$\text{source/opt/ros/kinetic/setup. bash}$$

Unless you add it to hidden '.bashrc' file in your home directory:

$$\text{echo "source/opt/ros/indigo/setup. bash" >> ~/. bashrc}$$

Master: Before starting with ROS, we must choose one of the devices in the ROS network as a master. It can be a robot computer (Turtlebot2) or a remote computer (Turtlebot3). The ROS master works as an intermediate node that aids connections between different ROS nodes. The master has all the details about all nodes running

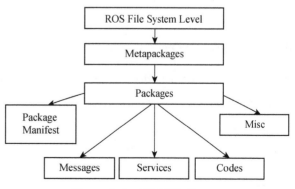

Figure 3-1 ROS File System

in the ROS environment.

ROS commands to navigate ROS file system:

rospack find [package_name]

roscd [locationname[/subdir]]

echo $ROS_PACKAGE_PATH

roscd log

rosls [locationname[/subdir]]

rospack find [package_name]

To create a ROS workstation:

mkdir -p ~/catkin_ws/src

cd ~/catkin_ws/src

catkin_init_workspace

cd ~/catkin_ws/

catkin_make

source devel/setup.bash

Testing:

echo $ROS_PACKAGE_PATH

To create a ROS package:

sudo apt-get install ros-<distro>-ros-tutorials

cd ~/catkin_ws/src

catkin_create_pkg beginner_tutorials std_msgs rospy roscpp

The above will create a folder called 'beginner_tutorials' and contains two files: package.xml and CMakeLists.txt. These two files have been partially filled out with the information you gave catkin_create_pkg.

cd ~/catkin_ws

catkin_make

source ~/catkin_ws/devel/setup.bash

Package dependencies:

rospack depends1 beginner_tutorials

roscd beginner_tutorials

cat package. xml

rospack depends1 rospy

To start ROS, we need to run the "roscore" command. Then, roscore is the first thing we run whenever we want to work on our project.

When creating a ROS package, we usually discuss node, topic, and service.

Node

ROS nodes are a process that uses ROS APIs to communicate with each other. A robot may have many nodes to perform its tasks. For example, an autonomous mobile robot may have a node for hardware interfacing, reading 2D lidar, and localization and mapping.

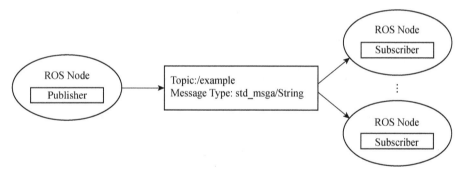

Figure 3-2   ROS Nodes

A node is only an executable file within a ROS package. ROS nodes use a ROS client library to communicate with other nodes.

rosnode list

rosnode info/rosout

rosnode ping my_turtle

rosrun command is used to run ros packages:

rosrun [package_name] [node_name]

rosrun turtlesim turtlesim_node

rosrun turtlesim turtlesim_node __name: = my_turtle

Topic

One of the methods to communicate and exchange ROS messages between two ROS nodes is ROS topics. Topics are named channels in which data is exchanged using ROS messages.

The turtlesim_node and the turtle_teleop_key node are communicating with each other over a ROS Topic. turtle_teleop_key publishing the critical strokes on a topic, while turtlesim subscribes to the same topic to receive the key strokes.

The command can display available topics and information about certaintopics is rostopic.

rostopic bw	display bandwidte used by topic
rostopic echo	print messages to screen
rostopic hz	display publishing rate of topic
rostopic lise	print information about active topics
rostopic pub	publish data to topic
rostopic type	print topic type

Example:

roscore

rosrun turtlesim turtlesim_node

rosrun turtlesim turtle_teleop_key

rostopic echo/turtle1/cmd_vel

rostopic list -h

Service

Services are another kind of communication method, like topics. Topics used to publish or subscribe to interactions, but in services are request or reply methods. One node can act as the service provider, which has a service routine running, and a client node requests a service from the server.

ROS tools

There are many tools in ROS provided, such as

rviz is one of the 3D visualizers available in ROS to visualize 2D and 3D values from ROS topics and parameters.

rqt_plot is a tool for plotting scalar values in the form of ROS topics.

rqt_graph, the ROS GUI tool can visualize the interconnection graph between ROS nodes.

Gazebo is an open-source dynamic robotic simulator with multiple robot models and extensive sensor support. The functionalities of Gazebo can be added via plugins.

Rosserial

Interfacing a microcontroller such as an Arduino board with ROS means running a ROS node, which can publish/subscribe like any ROS node on the microcontroller. For instance, an Arduino ROS node can be used to acquire and publish sensor values to a ROS environment, and other nodes can process it. Also, we can control devices, for example, actuators such as DC motors, by publishing values to an Arduino node. The primary communication between the PC and Arduino happens over UART. There is a dedicated protocol called ROS Serial.

ROS message

Sending ROS messages makes communications on 'topics' between nodes. For the publisher (turtle_tleop_key) and subscriber (turtlesim_node) to communicate, the publisher and subscriber must send and receive the same type of message.

We can look at the details of a message using rosmsg:

e. g. rosmsg show geometry_msgs/Twist

some examples about using rostopic command

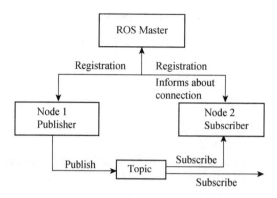

Figure 3-3　ROS Communication

rostopic type [topic], rostopic list, rostopic echo [topic]

e. g. rostopic type/turtle1/cmd_vel

To publish a message with a new topic:

rostopic pub [topic] [msg_type] [args]

e. g.

rostopic pub -1/turtle1/cmd_vel geometry_msgs/Twist — '[2.0, 0.0, 0.0]' '[0.0, 0.0, 1.8]'

rostopic pub/turtle1/cmd_vel geometry_msgs/Twist -r 1 — '[2.0, 0.0, 0.0]' '[0.0, 0.0, -1.8]'

To plot

rostopic hz [topic]

e. g. rostopic hz/turtle1/pose

Using rqt_plot

rosrun rqt_plot rqt_plot

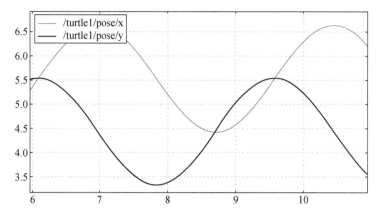

Figure 3-4 Using Rqt Plot Tools

Manipulators- II

In robotics, a manipulator can be defined as a device used to manipulate materials without direct interaction by its operator. Manipulators rooted in the automation stage of the industrial revolution to improve the production process. In addition, manipulators are used in applications where human safety is concerned, such as working with materials to be dealt with directly by a human, including radioactive or explosive materials. Recent developments have been used in various applications, including welding automation, robotic surgery, and space. Manipulators are one of the most crucial robotics divisions and one of the most desirable devices in some industrial production, such as the car industry.

The study of robot manipulators involves dealing with the positions and orientations of the several segments that make up the manipulators.

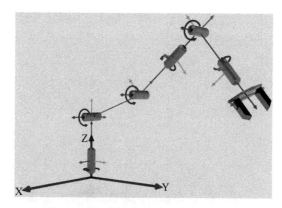

Figure 3-5 Manipulator with Gripper

Manipulators consist of an assembly of joints and links. In general, there are different types of joints:

Table 3-1 Different Types of Joints

Type	Figure	Description
Linear Joint	Input link — Output link	performs both translational and sliding movements
Orthogonal Joint	Input link — Output link	similar to linear joints, but input links will be moving at the right angles
Rotational Joint	Input link — Output link	rotary motion along the axis, vertical to the arm axes
Twisting Joint	Input link — Output link	twisting motion among the output and input links
Revolving Joint	Input link — Output link	the output link axis is perpendicular to the rotational axis, and the input link is parallel to the rotational axes

Degree of freedom:

In manipulators, the number of degrees of freedom of a mechanism is defined as the number

Chapter 3 Robotics Design and Control

of independent variables required to identify its configuration in space completely. The number of degrees of freedom for a manipulator can be calculated as:

$$n_{dof} = \lambda(n-1) - \sum_{i=1}^{k}(\lambda - f_i)$$

Where:

n is the number of links.

k is the number of joints.

f_i is the number of degrees of freedom of the ith joint.

λ is 3 for planar mechanisms and 6 for spatial mechanisms.

A 3×1 position vector can describe the position of any point in space which is relative to a reference frame. For instance, the position of point P concerning frame F can be displayed by a matrix:

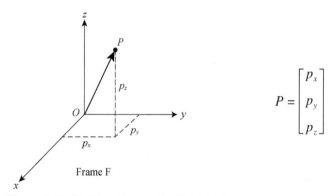

Figure 3-6 Position Vector of a Point in Space

Calculating Forward Kinematics

Forward kinematics is the task of calculating the position of the end effector from a given joint configuration (a set of joint angles).

In research, calculating the forward kinematics is the first step when using any new manipulator robot. No matter what software we use or what pre-define library for our programming language we have, the first step is to take a pen and paper to draw our robot joints' details.

The best starting point is to sketch all joints and links:

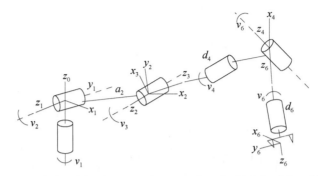

Figure 3-7　Using Pen and Paper to Sketch All Joints' Details

And a dynamic modelling of 4 dof robotic Arm:

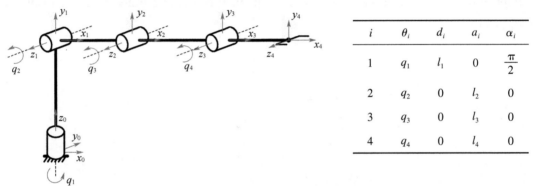

i	θ_i	d_i	a_i	α_i
1	q_1	l_1	0	$\frac{\pi}{2}$
2	q_2	0	l_2	0
3	q_3	0	l_3	0
4	q_4	0	l_4	0

Figure 3-8　Manipulator with 4 DOF

The DH approach method helps us to define joint parameters by assigning a different axis to each movable joint.

If we set up our axes correctly, working with the robot will be easy. Set them up incorrectly, and we will suffer countless headaches. Simulators and inverse kinematic solvers will require these axes.

Denavit-Hartenberg Method

To find the correct axis, we start from the first joint at the lowest part of the arm:

The Z-axis is the axis along the rotation or axis of translation for a prismatic joint.

X-axis, then, is for the frame and is a free choice.

The Y-axis is the axis constrained to complete the right-handed coordinate frame.

Chapter 3 Robotics Design and Control

For the second joint:

The X-axis is defined by the axis of the actuator.

X-axis is colinear with common normal, with origin at the intersection with new Z.

With these joint axes, four parameters define joint-to-joint transformation, d, θ, r, α.

d is the distance from the first joint's origin to the second joint's normal.

θ is the angle between the previous Z to align with the new X-axis.

r is the distance along the rotated X-axis.

α is the rotation about the new X-axis to move the previous z to the new one.

In joints with parallel Z axis, d is a free parameter and can be chosen freely, r is the distance between to axis of joints, θ calculated as before, and $\alpha = 0$.

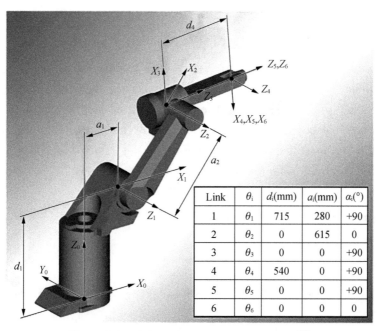

Link	θ_i	d_i(mm)	a_i(mm)	α_i(°)
1	θ_1	715	280	+90
2	θ_2	0	615	0
3	θ_3	0	0	+90
4	θ_4	540	0	+90
5	θ_5	0	0	+90
6	θ_6	0	0	0

Figure 3-9 Calculating Forward Kinematics[27]

The purpose of calculating forward kinematics is to get the end effector pose from the position of the joints. Today with many complicated grippers, the end effector pose of a manipulator's arm can be different[27].

Once we have determined parameters for all joints, time is to calculate forward kinematics.

$$T = \begin{bmatrix} & & & \vdots & \\ & R & & \vdots & T \\ & & & \vdots & \\ \cdots & \cdots & \cdots & \vdots & \cdots \\ 0 & 0 & 0 & \vdots & 1 \end{bmatrix} = \begin{bmatrix} \cos\theta & -\sin\theta\cos\alpha & \sin\theta\sin\alpha & \vdots & r\cos\theta \\ \sin\theta & \cos\theta\cos\alpha & -\cos\theta\sin\alpha & \vdots & r\sin\theta \\ 0 & \sin\alpha & \cos\alpha & \vdots & d \\ \cdots & \cdots & \cdots & \vdots & \cdots \\ 0 & 0 & 0 & \vdots & 1 \end{bmatrix}$$

For the right hand of the above equation, we create a 4×4 matrix for each arm joint up to the end effector and multiply it all together. T vector will contain the position of the end effector.

Once we have our D-H parameters, we can easily calculate forward kinematics using existing libraries in MATLAB Robotics Toolbox, ROS MoveIt, and many more.

Case Study: Robotic Arm with 4 DOF

Then by considering all joints a set, a joint configuration of a robotic arm is a set of all joint angles:

$$Q = \{\theta_1, \theta_2, \theta_3, \theta_4\}$$

It must be mentioned that the configuration space of a robotic arm is the area of all possible joint configurations "q" that it can have. Their dimension equals the number of joints or degrees of freedom—in our case, four. Therefore, the configuration space is constrained to a subset of all possible combinations of angles. Based on particular design or operational limits, each joint can move between two angles. Our case study is a robotic arm design in our robotics and AI lab with the following joint angles limitations:

θ_1: [−90, +90]   (Joint 1)
θ_2: [−25, +155]   (Joint 2)
θ_3: [−135, +45]   (Joint 3)
θ_4: [−90, +90]   (Joint 4)

Also, to define 'operational space' in our model, 'operational space' is a set of all possible end factor poses.

Obtaining Transformation Matrix

To obtain transformation matrixes, in the Denavit-Hartenberg convention, a total of four parameters are used to describe the transformation between two consecutive frames: two rotational parameters ("θ" and "α") and two linear parameters ("r" and "d"). Therefore, using D-H parameters, we have:

Chapter 3 Robotics Design and Control

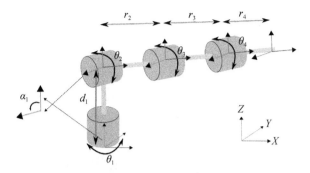

Joint	d(cm)	r(cm)	α(degree)	θ(degree)
J1	16.4	0	90	θ_1
J2	0	20	0	θ_2
J3	0	17	0	θ_3
J4	0	14.8	0	θ_4

Then the transformation matrix:

$$^{n-1}T_n = \begin{pmatrix} \cos\theta_n & -\sin\theta_n \cos\alpha_n & \sin\theta_n \sin\alpha_n & r_n \cos\theta_n \\ \sin\theta_n & \cos\theta_n \cos\alpha_n & -\cos\theta_n \sin\alpha_n & r_n \sin\theta_n \\ 0 & \sin\alpha_n & \cos\alpha_n & d_n \\ 0 & 0 & 0 & 1 \end{pmatrix} = \begin{pmatrix} R & T \\ 0\ 0\ 0 & 1 \end{pmatrix}$$

$^{n-1}T_n$ is the transformation from frame $(n-1)$ to frame (n);

R is rotation component;

And T is translation component.

By multiplying all transformation matrixes, we obtain a homogeneous transformation matrix calculated to describe the end effector's pose relative to the robotic arm's base.

$$\check{r} = 1T5 = 1T2 \times 2T3 \times 3T4 \times 4T5$$

The resulting matrix from the above gives us the position of the tip of the end effector concerning the body frame (coordinate frame).

Project: Arm Controller

Robotic Arm Controller
(2021−2022)

Robotic arm controller is a software developed in the robotics and AI labs of HYIT to control any manipulator from two to 6 dof for education purposes. All the parts, including the body, are developed in our lab. The application can control any combination of stepper and servo motors in serial communication and record and play functions. It is highly stable and accurate to 0.5 degrees in 50 times repetition. The package includes forward and inverse kinematics as well as D-H analysis.

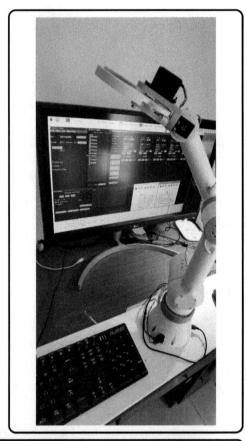

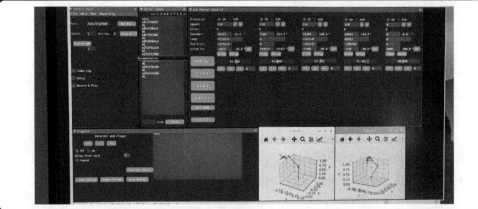

Chapter 3 Robotics Design and Control

Ground Moving Robots

Ground moving robots are divided to many different types.

Wheeled Robots

Wheeled robots navigate around the ground using motorized wheels to propel themselves. Wheel robots are widely used in many applications. They use less power, deal with locomotion mechanisms more easily, and perform faster. However, their operational environment is limited. Moving on uneven ground, stairs, and high slope requires a specific design to operate in such environments. There are many mechanically different wheeled robots for different functionality. Figure 3-10 displays some of the more popular designs.

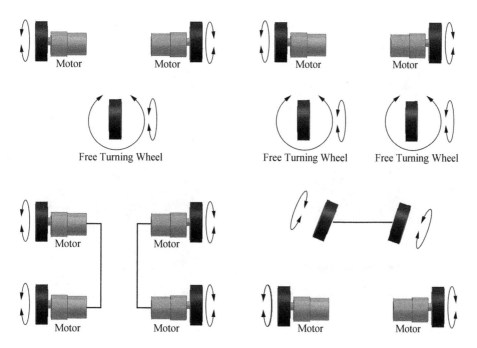

Figure 3-10 From Left to Right (differentially steered 3 wheeled, 2 powered- 2 free rotating, 4-wheel drive, differential steering)

In education, the choice even is more limited as the target is to provide experiments for theoretical knowledge with minimum design involvement. Fortunately, during the last decade, Turtlebot has gradually opened its way through robotics education with Open Source, ROS-based affordable robots. Turtlebot is a low-cost educational robot kit with Open-Source software packages created by Willow Garage in 2010. It comprises a robot base, a 3D sensor, and a computer or SBC (Single Board Computer).

The first series, Turtlebot1, consists of iRobot Create base, a 3000mAh battery pack, a Kinect sensor, and a laptop. Turtlebot2 has moved to a more powerful robot base, Yujin Kobuki, but with a smaller battery pack. Turtlebot2, in contrast, is more customisable and uses ROS1 (Robot Operating System). Turtlebot3, in collaboration with Robotis, became the smallest and cheapest of its generation. Turtlebot3 was a breakthrough without using a ready-made base. The base comprises a powerful development board, OpenCR and Dynamixel servo motors.

Turtlebot 1

Figure 3-11　Turtlebot-1　　　　Figure 3-12　Turtlebot-2

Turtlebot 2

Functional Specification

- Maximum translational velocity: 70 cm/s
- Maximum rotational velocity: 180 deg/s (>110 deg/s gyro performance will degrade)
- Payload: 5 kg (hard floor), 4 kg (carpet)

Chapter 3 Robotics Design and Control

- Cliff: will not drive off a cliff with a depth greater than 5cm
- Threshold climbing: climbs thresholds of 12 mm or lower
- Rug climbing: climbs rugs of 12 mm or lower
- Expected operating time: 3/7 hours (small/large battery)
- Expected charging time: 1.5/2.6 hours (small/large battery)
- Docking: within a 2 m × 5 m area in front of the docking station

Hardware Specification

- PC connection: USB or via RX/TX pins on the parallel port
- Motor overload detection: disables power on detecting high current (>3A)
- Odometry: 52 ticks/enc rev, 2578.33 ticks/wheel rev, 11.7 ticks/mm
- Gyro: factory calibrated, 1 axis (110 deg/s)
- Bumpers: left, centre, right
- Cliff sensors: left, centre, right
- Wheel drop sensor: left, right
- Power connectors: 5V/1A, 12V/1.5A, 12V/5A
- Expansion pins: 3.3V/1A, 5V/1A, 4 × analog in, 4 × digital in, 4 × digital out
- Audio: several programmable beep sequences
- Programmable LED: 2 × two-coloured LED
- State LED: 1 × two coloured LED [Green-high, Orange-low, Green & Blinking-charging]
- Buttons: 3 × touch buttons
- Battery: Lithium-Ion, 14.8V, 2200 mAh (4S1P-small), 4400 mAh (4S2P-large)
- Firmware upgradeable: via usb
- Sensor data rate: 50Hz
- Recharging adapter: Input: 100−240V AC, 50/60Hz, 1.5A max; Output: 19V DC, 3.16A
- Netbook recharging connector (only enabled when robot is recharging): 19V/2.1A DC
- Docking IR receiver: left, centre, right
- Diameter: 351.5mm/Height: 124.8mm/Weight: 2.35kg (4S1P-small)

Software Specification

- C++ drivers for Linux and windows
- ROS node
- Gazebo Simulation

Turtlebot start-up:

Turtlebot2 using ROS Indigo. After complete PC installation, a network setting is an essential first step:

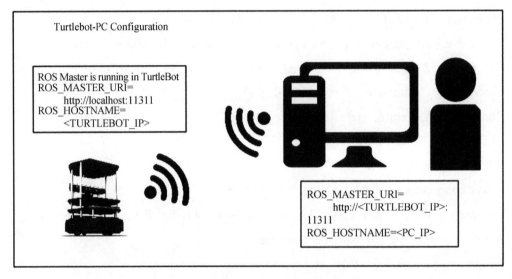

Figure 3-13　Turtlebot 2 Network Setup[28]

Network setup must be entered into '.bashrc' file. Using the command below on the robot:

❖ echo export ROS_MASTER_URI = http://localhost:11311 >> ~/.bashrc

❖ echo export ROS_HOSTNAME = IP_OF_TURTLEBOT >> ~/.bashrc

And on the remote PC:

❖ echo export ROS_MASTER_URI = http://IP_OF_TURTLEBOT:11311 >> ~/.bashrc

❖ echo export ROS_HOSTNAME = IP_OF_PC >> ~/.bashrc

One of the very noticeable differences between turtlebot2 and turtlebot3 is the 'master'. In turtlebot2, master is the robot, whereas in turtlebot3, master is the remote computer.

Chapter 3 Robotics Design and Control

To check correct network setup, we can run:

❖ Rostopic list

The above command should display all topics publishing from kobuki (ROS Master URI) base.

Also, we can publish a message on the remote PC:

❖ rostopic pub -r10 /hello std_msgs/String "hello"

and, on the turtlebot PC:

❖ rostopic echo /hello

The above command will display 'hello' message 10 times.

Clock synchronization is essential for ROS. Chrony is the best NTP client over lossy wireless. It needs clock synchronization if a robot behaves strangely when messages are sent from a PC application (like rviz, rqt, or ros node running on PC).

To install chrony

❖ sudo apt-get install chrony

manually sync NTP

❖ sudo ntpdate ntp.ubuntu.com

Bringup the robot is the first step in starting the robot. Following command should be executed on the robot PC:

❖ sudo ntpdate ntp.ubuntu.com

In the case of error, referring toturtlebot tutorial is helpful: http://wiki.ros.org/Robots/TurtleBot?distro = indigo.

Similarly, we need to start the remote PC too:

❖ rqt -s kobuki_dashboard

However, ifkobuki_dashboard is not installed, then:

❖ sudo apt-get install ros-indigo-kobuki-dashboard

running the command below will start visualization of the robot in RVIZ application. This is a standalone launcher, and doesn't need a connection to a running turtlebot. To view the default Turtlebot2 configuration:

* `roslaunch turtlebot_rviz_launchers view_model.launch-screen`

To launch RVIZ with the robot already running bringup, and a correct network setup:

* `roslaunch turtlebot_rviz_launchers view_robot.launch-screen`

Another useful command is 3D visualisation which is require for many tasks including SLAM and mapping:

After running 'minimal bringup' on the robot PC and correct configuration of the network:

* `roslaunch turtlebot_bringup 3dsensor.launch`

On the remote PC, startr viz already configured to visualize the robot and its sensor's output:

* `roslaunch turtlebot_rviz_launchers view_robot.launch`

Keyboard teleop is one of the ways we can interact with the robot. On the remote PC type:

* `roslaunch turtlebot_teleop keyboard_teleop.launch-screen`

or Joystick teleop:

1—PlayStation 3

* `roslaunch turtlebot_teleop ps3_teleop.launch-screen`

2—Xbox360 (use the left stick while keeping the right stick pressed in)

* `roslaunch turtlebot_teleop xbox360_teleop.launch -screen`

3—Logitech joysticks (general configuration for all logitech joysticks)

* `roslaunch turtlebot_teleop logitech.launch -scree`

Another very useful GUI application is Remocon:

Chapter 3 Robotics Design and Control

Figure 3-14 Turtlebot Interactions Window

Another useful application for remote controlling the robot is QT teleop:

```
sudo apt-get install ros-indigo-turtlebot-apps
sudo apt-get install ros-indigo-rocon-qt-teleop
```

Start theremocon and fire up the PC Pairing/Qt Joystick Teleop interaction.

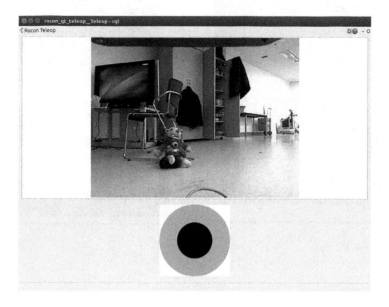

机器人与人工智能：智能机器人的实践与开发
Robotics and AI: Building Practical Intelligent Robots

For further information on operating your Turtlebot2, refer to tutorials at: http://wiki.ros.org/Robots/TurtleBot?distro = indigo.

Project: RIMA TX1

RIMA TX1-2016

RIMA TX1 was a complete indoor robot built in our robotics lab based on Turtlebot 2. There were three main parts of the design.
1—Body with raising neck, and arm
2—Voice recognition and text to speech
3—Chatbot development

Kobuki base
Kinnect 3D camera
Jetson TX1
ReSpeaker Core
Arduino Mega
I5 quad-core Laptop
CMOS camera

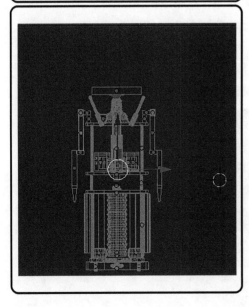

Chapter 3 Robotics Design and Control

Rima TX1 is a ROS (Robot Operating System) based mobile robot specially built for research in robotics. The hardware consists of one main four-core Intel i5 computer with a Linux operating system, one super microcontroller Nvidia Jetson Tx1, and three Arduino microcontrollers. Mapping and cloud point-based application, face recognition is supported with ASUS 3D pro live camera and a monocular camera. The Kabuki robot base is used for movement and ROS integrations. Six servo motors are used for arms and one for head movements. Also, an extendable neck is operated using a Bi-Polar stepper motor and a driver.

An Onboard 7" LCD touch-screen and a miniature keyboard/mouse are used when a remote connection is unavailable.

Rima TX1 uses a variety of sensors to provide extensive environmental information. Those include Infra-Red, Ultrasonic, 2D Laser Scanner, odometer, and bumper.

The primary platform selected for Rima TX1 is Linux Ubuntu ver. Ubuntu is an open-source Linux flavour based on Debian maintained and receives support from Canonical Ltd. Ubuntu currently is the most popular GNU/Linux operating system. Version 14.04 of Ubuntu is considered to be more reliable and robust for running ROS. The Robot Operating System (ROS) is a flexible framework for writing robot software. It is a collection of tools, libraries, and conventions that aim to simplify the creation of complex and robust robot behaviour across various robotic platforms. Over the past several years, ROS has grown to include a large community of users worldwide. Historically, most of the users were in research labs, but we increasingly see adoption in the commercial sector, particularly in industrial and service robotics.

We choose to use ROS indigo as some devices, such as Kabuki Robot base, ROS Kinetic and later, are not supported. Also, ROS is best tested on Ubuntu Linux.

Also another challenge was implementing ROS in Nvidia Jetson TX1. ROS for ARM processors is just a new work, and ARM Processor only partially supports much-related software such as Gazebo (simulation software).

We used ROS-Serial to communicate with the ROS network to implement various parts of Rima TX1, such as motors and sensors. These exclude various programs written for Arduino microcontrollers to control or acquire data from those parts.

All devices are orchestrated in ROS Operating System Network and communicate through Nodes with publishing messages.

Turtlebot3

Table 3-2 Turtlebot3 Specifications

Items	Burger	Waffle Pi
Maximum translational velocity	0.22 m/s	0.26 m/s
Maximum rotational velocity	2.84 rad/s (162.72deg/s)	1.82 rad/s (104.27deg/s)
Maximum payload	15 kg	30 kg
Size (L × W × H)	138 mm × 178 mm × 192 mm	281 mm × 306 mm × 141 mm
Weight (+ SBC + Battery + Sensors)	1 kg	1.8 kg
Threshold of climbing	10 mm or lower	10 mm or lower
Expected operating time	2h 30m	2h
Expected charging time	2h 30m	2h 30m
SBC (Single Board Computer)	Raspberry Pi	Raspberry Pi
MCU	32-bit ARM Cortex®-M7 with FPU (216 MHz, 462 DMIPS)	32-bit ARM Cortex®-M7 with FPU (216 MHz, 462 DMIPS)
Remote Controller	—	RC-100B + BT-410 Set (Bluetooth 4, BLE)
Actuator	XL430-W250	XM430-W210
LDS(Laser Distance Sensor)	360 Laser Distance Sensor LDS-01 or LDS-02	360 Laser Distance Sensor LDS-01 or LDS-02
Camera	—	Raspberry Pi Camera Module v2.1
IMU	Gyroscope 3 Axis Accelerometer 3 Axis	Gyroscope 3 Axis Accelerometer 3 Axis
Power connectors	3.3 V/800 mA 5V/4A 12V/1A	3.3 V/800 mA 5V/4A 12V/1A
Expansion pins	GPIO 18 pins Arduino 32 pins	GPIO 18 pins Arduino 32 pins
Peripheral	UART ×3, CAN ×1, SPI ×1, I2C ×1, ADC ×5, 5pin OLLO ×4	UART ×3, CAN ×1, SPI ×1, I2C ×1, ADC ×5, 5pin OLLO ×4
DYNAMIXEL ports	RS485 ×3, TTL ×3	RS485 ×3, TTL ×3

Chapter 3 Robotics Design and Control

(Continued)

Items	Burger	Waffle Pi
Audio	Several programmable beep sequences	Several programmable beep sequences
Programmable LEDs	User LED × 4	User LED × 4
Status LEDs	Board status LED × 1 Arduino LED × 1 Power LED × 1	Board status LED × 1 Arduino LED × 1 Power LED × 1
Buttons and Switches	Push buttons × 2, Reset button × 1, Dip switch × 2	Push buttons × 2, Reset button × 1, Dip switch × 2
Battery	Lithium polymer 11.1V 1800 mAh/19.98Wh 5C	Lithium polymer 11.1V 1800 mAh/19.98Wh 5C
PC connection	USB	USB
Firmware upgrade	via USB/via JTAG	via USB/via JTAG
Power adapter (SMPS)	Input: 100-240V, AC 50/60Hz, 1.5A @ max Output: 12V DC, 5A	Input: 100-240V, AC 50/60Hz, 1.5A @ max Output: 12V DC, 5A

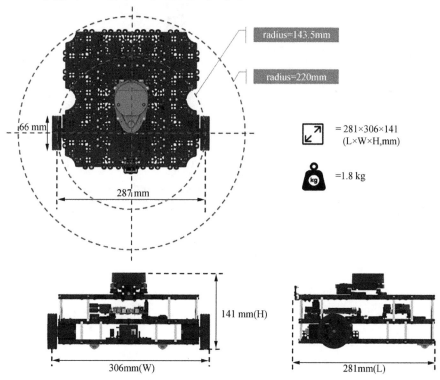

Figure 3-15 Turtlebot3 Waffle Pi

There is an online e-manual for Turtlebot3, provided by robotis. com at: https://emanual. robotis. com/docs/en/platform/turtlebot3/overview/.

Automated Guided Vehicle Systems (AGVS)

AGVS is the collaboration of sensors, software, and hardware that guides the intelligent navigation of a vehicle while performing its task or moving from one point to another. Typically, these vehicles, known as automated guided vehicles (AGVs), are battery-powered and use several guidance technologies, including lasers, optical sensors, gyroscope/magnet, magnetic tapes, or bars, making it easy to change routes and avoid obstacles in response to changes in the vehicle's immediate environment.

Figure 3-16 ROS Based AGV by Robotnik[14]

An AGV is any vehicle that can sense its environment and respond intelligently without human intervention. It is an automated vehicle that carries a load in an assembly, manufacturing line, or warehousing facility. An AGV can take on any form that best suits the users' needs. Engineers have researched and experimented with various iterations of self-driving vehicle prototypes for decades. Theoretically, the concept is simple—equip a vehicle with sensors that can track everything in its immediate radius and have the car respond intelligently.

AGVs have historically been used in industries to transport heavy material in and around large buildings such as factories or warehouses. In recent times, the applications of AGVs have become broader, giving rise to various categories and definitions for what an AGV and

Chapter 3　Robotics Design and Control

Figure 3-17　Ridgeback a ROS Based AGV From Clearpath[29]

AGVS are. In the 1950s, Arthur M. Barret Jr invented the first AGV—a modified towing tractor configured to follow an overhead wire and used to pull a trailer in a grocery warehouse. It was soon followed by a rudimentary AGV introduced into the market by Barret Electronics of Northbrook, Illinois—a tow truck that followed a wire embedded into a warehouse or factory floor. AGVs and AGVS are constantly being improved in safety, navigation, power, task complexity, control units, and systems, among others; these innovations are made possible by advancements in electronics, computer technology, programming, and sensor technology.

Humanoid Robots

This design of wheeled robots is more straightforward than treads or legs, and using wheels makes it easier to design, build, and program for movement in flat, not-so-rugged terrain. However, their operation is more limited. On the one hand, designing a humanoid robot, especially for computer science and software engineering students, is a big challenge; on the other hand, there are many research areas in which students need to use a humanoid robot.

Nao is one of the most famous robots for research and education. With 25 degrees of freedom, Nao is an autonomous, programmable humanoid robot developed by Aldebaran Robotics, a French robotics company headquartered in Paris, which was acquired by SoftBank Group in 2015 and rebranded as SoftBank Robotics.

机器人与人工智能：智能机器人的实践与开发
Robotics and AI: Building Practical Intelligent Robots

Figure 3-18　NAO Robot

NAO is not cheap, but still more affordable than other advance humanoid robots. The sixth-generation NAO enables unique human-robot interaction. It is the most advanced humanoid, and its applications for teaching programming are endless! In the classroom, the V6 helps teach coding, brings literature to life, enhances special education, and allows training simulations. Its educational solutions include an intuitive interface, remote learning, and various applications for accessibility! With the NAO V6, students can code in Java, Python, C++, C#, and more.

"With numerous upgrades, the NAO V6 is outfitted with the absolutely latest robotics technology. Its motherboard boasts 4 GB of RAM, a Quad Core CPU, and 32 GB of SSD HD while its on-board camera includes autofocus, a 68.2 DFOV field of view, and 30 cm–infinity focus range. The V6 also includes improved audio functions, including omnidirectional

Chapter 3   Robotics Design and Control

microphones and a loudspeaker as well as new motors with a lifetime improvement. With a faster boot time of 51 seconds, system upgrade time of under five minutes, Wi-Fi that's up to 10 times better than the V5, and Ethernet performance four times faster than the V5, this new model is poised to perfect robotics in any classroom! Speaking of faster, the V6's CPU is twice as effective as the V5 and its RAM is five times more effective than the V5! Plus, its read speed is 16 times faster than the V5 and write speed is 35 times faster, making the newest NAO a stalwart in any educational setting."[30]

Table 3-3   NAO Robot Specifications

	NAO Evolution	NAO⁶
Motherboard	ATOM Z530 **1.6 GHz** CPU **Bi core** **1 GB** RAM **2 GB** Flash memory + 8 GB Micro SDHC	ATOM E3845 **1.91 GHz** CPU **Quad core** **4 GB** DDR3 RAM **32 GB** SSD
Camera	Field of view : 72.6°DFOV (60.9°HFOV,47.6°VFOV) Focus range: 30cm – infinity Focus type : **Fixed focus**	Field of view : **68.2°**DFOV (57.2°HFOV,44.3°VFOV) Focus range : 30cm – infinity Focus type : **Autofocus**
Audio	Cardioid microphones (-12dB) Loudspeakers: Foxconn Amato	Omnidirectional microphones (-12dB) + Audio codec Loudspeakers: Seltech 40S19 custom
Motor		New motors with a lifetime improvement
Color	Red, blue, light grey	Dark grey

Quadruped Robots

Following Boston Dynamic Spot®, a quadruped robot, a new imagination of new generation robots started to grow both in industry and academia. For many, Spot was an unreachable dream with a cost of about 75,000 US dollars.

A1 is a quadruped robot capable of analysing its environment, monitoring and overcoming obstacles, thanks to its intelligent sensitization.

Its comprehensive mobility allows it to carry out many tasks: from parcel and material delivery tasks through environmental inspection and security applications (recognition in dangerous areas) or within the framework of entertainment activities. On the other hand, it is the ideal robot for the R&D sector, where the applications are endless.

Figure 3-19 A1 Quadruped Robot

- Higher running speed: Maximum continuous outdoor running speed at 3.3 m/s (11.88 km/h)
- Multi-eye intelligent camera: 1080 P resolution
- Excellent motion stability
- High strength and light body structure
- Integrated force sensor at each foot end
- Vision-based autonomous object tracking
- Realtime tracking of objects within visual range
- Avoid obstacles within 0.8 m of the robot's visual range
- High precision lidar to map building, autonomous positioning, navigation planning and dynamic obstacle avoidance
- Friendly user interface

A1 contains two single board computers. It also supports ROS.

Chapter 3 Robotics Design and Control

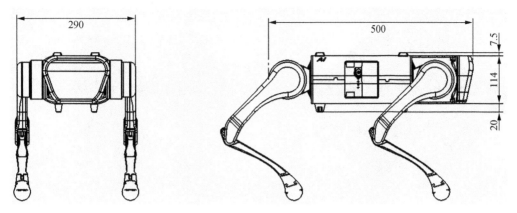

Figure 3-20 A1 Quadruped Robot Dimensions

Toward Intelligent Robotics

Intelligent Vision Sensors

Pixy Camera

The Pixy2 CMUcam5 is smaller, faster and more capable than the original Pixy. Like its predecessor, the Pixy2 can learn to detect objects we teach by pressing a button. The Pixy2 has new algorithms that detect and track lines for use with line-following robots. With these new algorithms, we can detect intersections and "road signs" as well. The road signs can tell our robot what to do, such as turn left, right, slow down, and more. The best part is that the Pixy2 does all of this at 60 frames per second, so our robot can be fast.

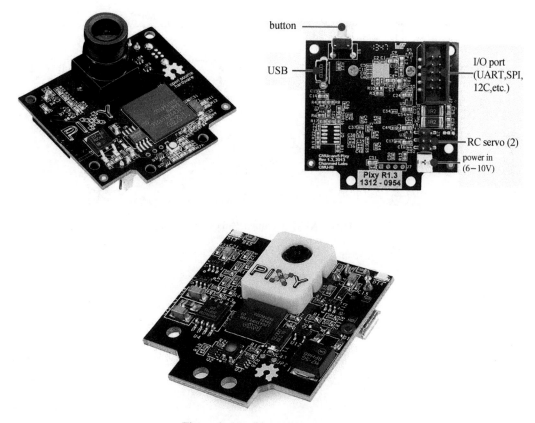

Figure 3-21　Pixy Camera Board

OpenMV Camera

The OpenMV Cam is a small, low-power, microcontroller board which allows us easily and directly to implement applications without additional microcontroller or SBC. It is using built-in machine vision in the real world. We can program the OpenMV Cam in high-level Python scripts. The OpenMV Cam has a cross-platform IDE (based on Qt Creator) explicitly designed to support programmable cameras.

Chapter 3 Robotics Design and Control

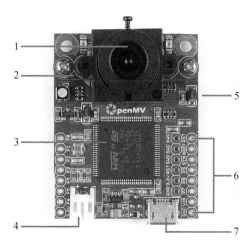

1—OpenMV camera using OV7725 image sensor;
2—Modular sensor design supports and multiple sensors;
3—32-Bit Arm Cortex-M7 operating at 400 MHz;
4—LiPo battery connector;
5—MicroSD card slot (supports up to 64GBs);
6—All I/O pins output 3.3 V and are 5 V tolerant;
7—Micro-USB full speed or programming.

Figure 3-22 OpenMV Camera Board

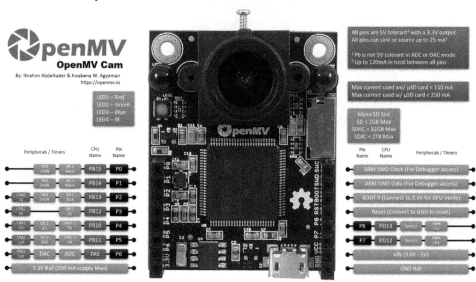

Support for TensorFlow

The OpenMV firmware also supports loading quantized TensorFlow Lite models. Also, the firmware supports loading external models that reside on the filesystem to memory (on boards with SDRAM) and internal models (which are embedded into the firmware) in place. If we need to load the external TensorFlow model from the filesystem from Python, we can use the tf Python module. For information on embedding TensorFlow models into the firmware and loading them, please see TensorFlow Support.

OpenCR DepthAI Cameras

OAK-1

The Oak-1 is an all-in-one machine vision solution, and it is a 4-trillion-operations-per-second AI powerhouse that performs our AI models onboard so that our host is free to do whatever we need.

Figure 3-23 OAK-1 Camera

OAK-D

The DepthAI hardware, firmware, and software suite combine depth perception, object detection (neural inference), and object tracking and give you this power in a simple, easy-to-use Python API. This OAK-D variant of DepthAI allows the power of the Myriad X VPU to be fully harnessed and used with the platform of our choice.

Chapter 3 Robotics Design and Control

Figure 3-24 OAK-D Camera

The OAK-D board includes three 7.5 cm stereo depth onboard cameras, allowing quick setup and interfaces over USB3C to the host, which allows DepthAI to be used with our (embedded) host platform of choice, including the Raspberry Pi 4, Jetson Nano Dev Kit, Google Edge TPU Coral, and other popular embedded hosts and operating systems.

OAK-D Lite

The OAK-D Lite has the main features of the OAK-D, making them even better. It is smaller and more affordable, and OAK-D Lite can still be considered with all those computer vision power. The main difference between the D and the D Lite are lower-resolution mono-cameras, no IMU on board, and no power adapter. For most use cases of the OAK-D Lite, the USB Type C port should supply enough power (a Y-adapter is available through Luxonis should extra power be needed).

Figure 3-25 OAK-D Lite

Smell Inspector (e-nose)

Finally, our robot can smell and learn and recognise newly trained smells using a smell Inspector.

Smell Inspector: E-Nose Developer's Kit and End-User Gadget developed by Smart

Nanotubes Technologies through a Kickstarter campaign.

Smell Inspector is a developer's kit and ready-to-use electronic nose gadget in one. For anyone working on intelligent systems—from developers or scientists to tinkerers. Immediately usable for manufacturers who want to control their production environment and the quality of their products.

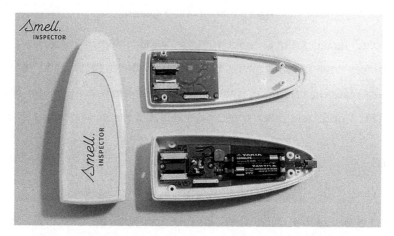

Figure 3-26　Smell Inspection Sensor Based on Machine Learning

A smell inspector is a digital nose, which works with the same principle as a human nose. Train it and create our specific use case for single gases, gas mixtures or smells.

Smell Sensor consists of an electronic board with four Smell iX16 detector chips, providing full functionality to read signals from all 64 individual detectors every 1.8 s and send them out in ASCII format. It can be integrated into different mobile or stationary appliances.

Smell iX16

Smell Sensor is the core invention: the first multi-channel gas detector chip designed to conquer the mass market. The sensor elements contain fine-tuned carbon nanomaterials, making the chip highly sensitive, small, energy-efficient and affordable. There are four Smell iX16 detector chips on the electronic board. The current version has 64 detectors (16 channels per detector chip) with a size of less than 1 sqmm (0.08 sqinch) each. Further miniaturization and/or increasing the number of detectors can quickly be done. The detector chip Smell iX16 only needs one μW of power supply. Various host devices, energy harvesters or batteries can thus power it. Smell iX16 is designed for unlimited miniaturized electronic and IoT applications in various areas. The detector chip has a high sensitivity to different gases and VOCs.

Smell Annotator

Smell Annotator is an AI-based software used to detect, annotate and digitize smells with the Smell Inspector and recognize annotated smells. 64 independent gas detectors produce digital smell patterns. This technology learns to recognize particular smells through artificial intelligence, which can be compared to how information is processed in our olfactory system. The smell is initially collected and analysed in the olfactory bulb. For recognition and remembering, the cortical brain areas get involved. This evolves over our whole life with more and more data processed. The same principle is realized in the Smell Inspector, supported by the Smell Annotator software, which uses machine learning algorithms. Currently, Windows and Linux systems are supported. Android and iOS will follow.

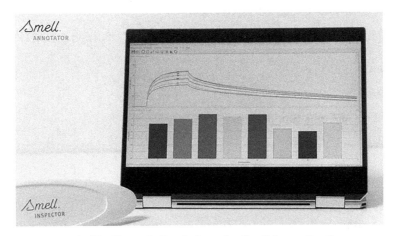

Figure 3-27　Example Result from the Smell Inspection Sensor

Smell Inspector Technical Details:

Dimensions: $166 \times 64 \times 31$ mm/$6.54 \times 2.52 \times 1.22$ inch ($L \times W \times H$)

Power consumption: max. 0.28 W

Weight: 130 g/0.29 lb

Materials: Smell Board iX16 ×4, housing

Smell Board iX16 ×4:

Dimensions: 157×40 mm and 67×50 mm/6.18×1.58 inch and 2.64×1.97 ($L \times W$)

Power consumption: max. 0.28 W

Weight: 45 g/0.1 lb

Materials: 4 Smell iX16 detector chips, 2 PCBs

Read-out time: all 64 channels every 1.8 s

Read-out format: ASCII

Serial interface

Compatibility: Compatible with Arduino and Raspberry Pi

Smell iX16:

Dimensions: 22×8 mm/0.87×0.32 inch ($L \times W$)

Power consumption: $1 \mu W$

Weight: 0.1 g/0.0002 lb

Materials: 16chemiresistor-type nanomaterial based gas detectors, Kapton foil, particle filter

Temperature: 0°C to +40°C/32°F to +104°F

Sensitivity: High sensitivity to different gases and VOCs (<80 ppb for NH_3, PH_3, H_2S)

Smell Annotator:

Smell recognition

Smell digitalization and annotation

Visualization of measurements

Saving measurements and loading saved data

Data format: CSV

Compatibility/Connectivity

"Maximum Flexibility in Smell Recognition and Connectivity"

Smell Annotator Environment

Creator's analytic software Smell Annotator is supplied with Smell Inspector. It can be used to view, annotate and store measurements of the Smell Inspector. In all cases, we will simultaneously get lifetime access to data from all 64 gas detector channels. Based on their pre-set data collection, Smell recognition will be available for a limited set of smells and gases. Smell Annotator is a development tool! Extend our smell data collection and teach our Smell Inspector to recognize new fragrances and their compositions! Currently,

Chapter 3 Robotics Design and Control

Windows and Linux systems are supported. Android and iOS will follow. Alternatively, we may run and test our AI-based software.

Exercises and Projects

Changes in the field of robotics are speedy. Therefore, to provide more UpToDate, more up-to-date problem-solving techniques and innovative projects, this part will be available from the author's website at www. roboticacamp. com/robotics_and_ai.

Chapter 4

Artificial Intelligence and Machine Learning

Introduction

"Machine learning" is one of the current technology buzzwords. Machine learning is often used parallel with artificial intelligence, deep learning, and big data, but what does it mean?

AI (Artificial Intelligence) can solve tasks which require human intelligence. At the same time, ML is a subset of artificial intelligence that solves specific tasks by learning from data and making predictions without being programmed explicitly for it, which means that all machine learning is AI, but not all AI is machine learning.

Artificial intelligence and machine learning are vast areas of study and require years of learning with a targeted direction. It does not mean we cannot use and benefit from AI with our robot project development. Using machine learning frameworks for various functionality made a fundamental shift in robotics development in the recent decade. These functionalities include face recognition, object recognition, voice recognition, and text and handwritten text recognition.

Also, there are a tremendous amount of information and learning materials, as well as cloud base machine learning ready machines provided by Amazon, Microsoft and others which makes learning and using machine learning simpler and more joyful.

Then, the purpose of this chapter becomes specific to introducing some terminologies, exploring deep learning and its frameworks, and finally, and most importantly, introducing devices and methods to make our robotics projects practically intelligent.

Chapter 4 Artificial Intelligence and Machine Learning

Terminology

What is an algorithm?

An algorithm is a set of rules a machine follows to achieve a particular goal. An algorithm can be considered a recipe that defines the inputs, the output and all the steps needed to get from the inputs to the output.

What is a dataset?

A dataset is a table with the data from which the machine learns. The dataset contains the features and the target to predict. When used to induce a model, the dataset is called training data.

What is an instant?

An instance is a row in the dataset. Other names for 'instance' are (data) point, example, and observation.

What is a feature?

The features are the inputs used for prediction or classification, and a feature is a column in the dataset.

What is AI?

According to the father of Artificial Intelligence, John McCarthy, it is "the science and engineering of making intelligent machines, brilliant computer programs"[31].

Artificial intelligence, is a computer ability or machine to mimic or imitate intelligent human behaviour and perform human-like tasks. Human intelligence requires thinking, reasoning, learning from experience, and finally, making decisions.

What is ML (Machine Learning)?

Machine learning belongs to the artificial intelligence domain, which can autonomously learn from the data without being directly controlled by a programmed or assisted by domain expertise.

What is deep learning?

Deep learning is one of the machine learning frameworks that is a set of algorithms inspired by the structure and function of the human brain neurons.

What is TensorFlow?

TensorFlow is the second machine learning framework that Google created and used to design, build, and train deep learning models.

We can use the TensorFlow library to do numerical computations, which does not seem too special, but these computations are done with data flow graphs.

CPU or GPU?

Deep learning can be experienced both with CPU and GPU. Of course, GPUs are much faster for practical applications. Similarly, we can use TensorFlow with CPU.

What Is Neural Network?

A neural network is a network of biological neurons. Whereas an artificial neural network is a series of algorithms that endeavours to recognize underlying relationships in a data set through a process that mimics how the human brain operates. Therefore, neural networks refer to systems of neurons, either organic or artificial.

A biological neural network comprises groups of neurons which chemically connected or functionally associated together. While a single neuron may be connected to many other neurons, a network's total number of neurons and connections may be extensive. "Connections, called synapses, are usually formed from axons to dendrites, though dendrodendritic synapses"[3] and other connections are possible. Apart from electrical signalling, other forms of signalling arise from neurotransmitter diffusion[32].

Artificial neural network is inspired from biological model, with neuron like nodes with inputs and outputs.

Chapter 4   Artificial Intelligence and Machine Learning

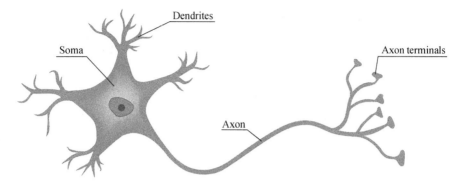

Figure 4-1   Biological Neuron[33]

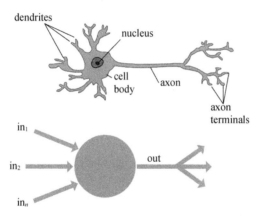

Figure 4-2   Human Brain Neuron

Neural networks can adapt to changing the input to generate the best possible result without redesigning the output criteria.

In the human brain, for a neuron to become active, the total input must reach a threshold at which excitation outweighs inhibition. Only at this point will the receiving neuron spike.

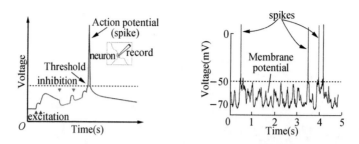

Figure 4-3 Action Potential (spike) in Neurons

In artificial neurons, similarly, the total input weights must reach a threshold to spike at the output(s).

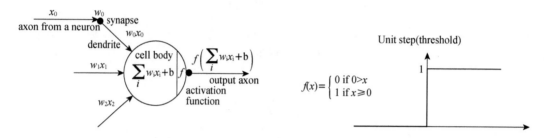

Figure 4-4 Artificial Neurons

When the total weighted inputs reach the threshold defined by the activation function, the output(s) spikes.

Perceptron

Perceptron was developed in the 1950s and 1960s by the scientist Frank Rosenblatt, inspired by earlier work by Warren McCulloch and Walter Pitts. The neuron's output, 0 or 1, determines whether the weighted sum $\sum_j w_j x_j$ is less than or greater than some threshold value.

$$\text{output} = \begin{cases} 0 & \text{if } \sum_j w_j x_j \leq \text{threshold} \\ 1 & \text{if } \sum_j w_j x_j > \text{threshold} \end{cases}$$

In machine learning, a perceptron is an algorithm used for supervised learning. A perceptron is an algorithm in a neural network that performs a rule base computation to detect features in the input data. Perceptron, also known as Linear Binary Classifier.

Chapter 4　Artificial Intelligence and Machine Learning

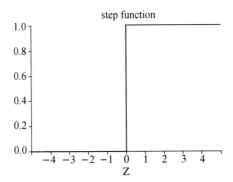

Figure 4-5　Perceptron's Function

In the example for having coffee, we can discuss three conditions:

1—If I'm thirsty;

2—If my friend invites me;

3—If I have had already a cup of coffee today.

The decision will be performed based on the weight of each condition as well as the threshold for the decision.

However, we can get different decision-making models by varying the weights and the threshold.

Perceptron is not a complete model of human decision-making; instead, a complex perceptron network could make quite subtle decisions.

The perceptron consists of the following:

Input: The input layer of the perceptron is artificial inputs neurons which enter the initial data for further processing.

Weights and Bias:

Weight: It represents the strength and influence of the connection between units. For instance, if the weight from one node to the next has a higher quantity, the neuron has a more considerable influence on the neuron.

Bias: The task to modify the output along with the weighted sum of the input to the other neuron.

Net sum: It represents the total sum.

Activation function: A neuron that can be activated or not is determined by an activation function. The activation function calculates a weighted sum and further adds bias to it to give the result.

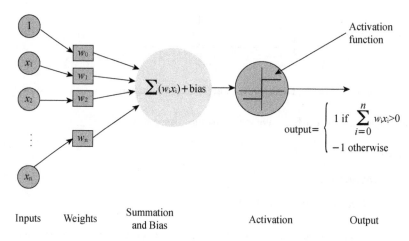

Figure 4-6 Perceptron Complete Model

There are two types of architecture of perceptron:

—Single Layer Perceptron

—Multi-Layer Perceptron

The first neural network proposed in 1958 by Frank Rosenbluth was the single layer perceptron, and it is one of the earliest models for learning. The goal was to find a linear decision function measured by the weight vector w and the bias parameter b.

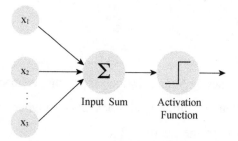

Figure 4-7 Single Layer Perceptron[34]

However, multiple hidden layers are used for the non-linearity of the data. This instruction is also called a feed-forward network.

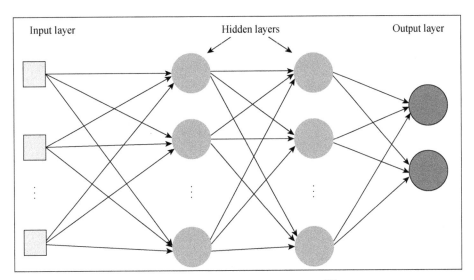

Figure 4-8　Multi-Layer Perceptron[35]

Sigmoid Neuron

Sigmoid neurons are similar to perceptron, but they are slightly modified where the output from the sigmoid neuron is much smoother than from the perceptron.

We should change some weight (or bias) in the network to see how learning might work. We want this small weight change to cause only a tiny corresponding change in the output from the network.

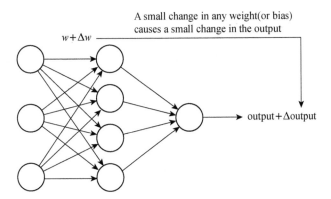

Figure 4-9　Changes in Weight Cause Changes in Output

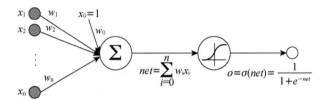

Figure 4-10　Sigmoid Neuron Model

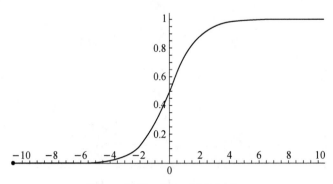

Figure 4-11　Sigmoid Function

The perceptron model takes several real-valued inputs and gives a single binary output. In the perceptron model, every input x_i weights w_i associated with it. The weights indicate the importance of input in the decision-making process. Also, perceptron output is rigorous. In contrast with the earlier example about having coffee, consider the famous example of the decision-making process of a person, whether he/she would like to purchase a car or not based on only one input x_1—Salary and by setting the threshold $b(W_0) = -10$ and the weight $w_1 = 0.2$. The output from the perceptron model will look like this in the figure below.

The smoothness of σ means that small changes Δw_j in the weights and Δb in the bias will produce a small change Δoutput in the output from the neuron. Calculus tells us that Δoutput is well approximated by

Where the sum is over all the weights, w_j, and $\partial \text{output}/\partial w_j$ and $\partial \text{output}/\partial b$ denote partial derivatives of the output with respect to w_j and b, respectively.

$$\Delta \text{output} \approx \sum^j \frac{\partial \text{output}}{\partial w_j} \Delta w_j + \frac{\partial \text{output}}{\partial b} \Delta b$$

A demonstration in python can explain sigmoid output as in the code below:

Chapter 4 Artificial Intelligence and Machine Learning

```
In [1]: from math import e

In [2]: def sigmoid(x):
            return 1/(1+e**-x)

In [3]: sigmoid(1)
Out[3]: 0.7310585786300049

In [4]: sigmoid(-100)
Out[4]: 3.7200759760208555e-44

In [5]: sigmoid(100)
Out[5]: 1.0
```

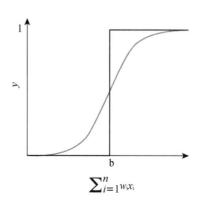

If the input is minimal, such as −100, then the output is close to zero. If the input is a large number, then the output is near '1.0'.

Activation Functions

There are many activation functions besides the step function used in Perceptron. When we have a regression problem to solve, the linear activation function is used. The binary step activation function is used in Perceptron. It cannot be used in multi-layer networks. The sigmoid activation function can be used both at the output layer and hidden layers of a multi-layer network.

Table 4-1 Common Activation Functions[36]

Name	Plot	Function, $f(x)$	Derivative of f, $f'(x)$	Range
Identity		x	1	$(-\infty, \infty)$
Binary step		$\begin{cases} 0 & \text{if } x < 0 \\ 1 & \text{if } x \geq 0 \end{cases}$	$\begin{cases} 0 & \text{if } x \neq 0 \\ \text{undefined} & \text{if } x = 0 \end{cases}$	$\{0, 1\}$
Logistic, sigmoid, or soft step		$\sigma(x) = \dfrac{1}{1+e^{-x}}$ [1]	$f(x)(1-f(x))$	$(0, 1)$
tanh		$\tanh(x) = \dfrac{e^x - e^{-x}}{e^x + e^{-x}}$	$1 - f(x)^2$	$(-1, 1)$
Rectified linear unit (ReLU)[11]		$\begin{cases} 0 & \text{if } x \leq 0 \\ x & \text{if } x > 0 \end{cases}$ $= \max\{0, x\} = x\mathbf{1}_{x>0}$	$\begin{cases} 0 & \text{if } x < 0 \\ 1 & \text{if } x > 0 \\ \text{undefined} & \text{if } x = 0 \end{cases}$	$[0, \infty)$

Cost Function

The importance of the cost function is in measuring "how good" a neural network did concerning its given training sample and the expected output.

Also, the cost function may depend on variables such as weights and biases. A cost function is a single value, not a vector, because it is value represents how well the neural network worked as a whole.

$$C(w, b) \equiv \frac{1}{2n} \sum_{x} \| y(x) - a \|^2.$$

Here, w denotes the collection of all weights in the network, b is all the biases, n is the total number of training inputs, a is the vector of outputs from the network when x is input, and the sum is over all training inputs, x.

$C(w, b)$ is non-negative since every term in the sum is non-negative.

The cost $C(w, b)$ becomes small, i.e. $C(w, b) \approx 0$, precisely when $y(x)$ is approximately equal to the output, a, for all training inputs, x.

So our training algorithm has done an excellent job if it can find weights and biases so that $C(w, b) \approx 0$.

If we instead use a smooth cost function like the quadratic cost, it is easy to figure out how to make small changes in the weights and biases to improve the cost.

We focus first on minimizing the quadratic cost, and only after that will we examine the classification accuracy.

What we would like is to find where C achieves its global minimum. Now, of course, for the function plotted (right), we can eyeball the graph and find the minimum.

A general function, C, maybe a complicated function of many variables, and it will not usually be possible just to eyeball the graph to find the minimum.

Chapter 4 Artificial Intelligence and Machine Learning

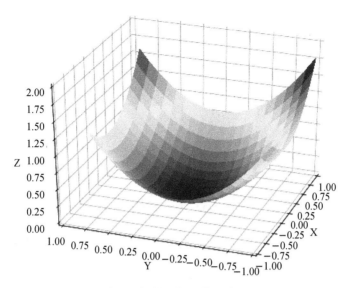

Figure 4-12 Cost Functions

Gradient Descent

Gradient descent is an iterative algorithm used to find a local minimum/maximum of a given function. This method of calculating minimum/maximum is commonly used in machine learning or deep learning to minimise a cost or loss function.

$$\Delta C \approx \frac{\partial C}{\partial v_1}\Delta v_1 + \frac{\partial C}{\partial v_2}\Delta v_2.$$

Sigmoid is a family of functions where we can get various sigmoid functions by changing the values of 'w' and 'b'.

Changing values of 'w' and 'the' slope and the curve's displacement can change.

We are given input and output, and we have chosen to approximate using the Sigmoid function with parameters 'w' and 'b'.

Also, the gradient descent algorithm does not work for all functions. Two specific requirements are differentiable and convex a function has to be.

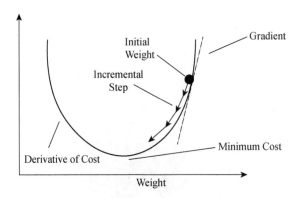

Figure 4-13 Gradient Descent

Let us think about what happens when we move the ball a small amount Δv_1 in the v_1 direction and a small amount Δv_2 in the v_2 direction. Calculus tells us that C changes as follows:

We will find a way of choosing Δv_1 and Δv_2 to make ΔC negative; i.e. we will choose them, so the ball is rolling down into the valley.

The way the gradient descent algorithm works is to compute the gradient ΔC repeatedly and then move in the opposite direction, "falling" the valley's slope. We can visualize it like this: [37]

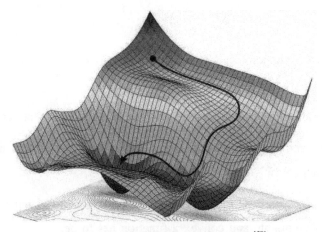

Figure 4-14 Gradient-based Learning[38]

Imagine if we were on the side of a mountain with no map and needed help to reach the top. Then we decided to take one step, going further in the direction of the steepest and easiest to

Chapter 4 Artificial Intelligence and Machine Learning

go up the current position, and then continuing to take a small step in the steepest direction of the next position. Step by step, we went all the way to think that we have reached the top of the mountain. Here the direction of the mountain through the steepest path is the gradient.

Cost Functions for Evaluating Performance of Neural Network

A feedforward neural network is many layers of neurons connected. It is often called feedforward neural networks or multi-layer perceptron (MLPs). Then it takes in an input that "trickles" through the network, and then the network returns an output vector.

More formally, call a_j^i the activation (aka output) of the jth neuron in the ith layer, where a_j^1 is the jth element in the input vector.

Then we can relate the next layer's input to its previous via the following relation:

$$a_j^i = \sigma\left[\sum_k (w_{jk}^i) \cdot a_k^{i-1}\right] + b_j^i$$

σ is the activation function,
w_{jk}^i is the weight from the Kth neuron in the $(i-1)^{th}$ layer to the jth neuron in the ith layer, b_j^i is the bias of the jth neuron in the ith layer, and a_j^i represents the activation value of the jth neuron in the ith layer.

Usually, we written as z_j^i to represent $\sum_k (w_{jk}^i \cdot a_k^{i-1}) + b_j^i$, in other words, the activation value of a neuron before applying the activation function.

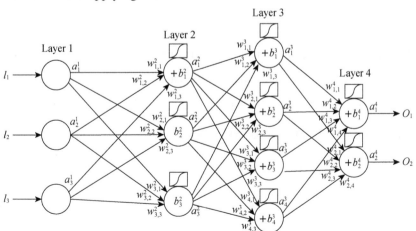

Figure 4-15 Activation Function of Feedforward Neural Network

Example Use Case of DNN

Deep learning for hearing aid devices

"DNN was trained with millions of real-life sound scenes—a restaurant, train station or busy street. The DNN would learn to identify and balance each sound within it, so you can access the sounds most important to you. So that's what exactly has been done! They trained a DNN with 12 million complex real-life sound scenes like these, which it then learned to analyse, organize and balance. After it had learned all this awesome knowledge, it was ready to power the hearing device. Now this hearing device can utilize the DNN's intelligent capabilities when balancing and prioritizing the sounds that are important to you, which also supports your brain health"[39].

Use of deep learning in recognizing hand-written numbers

Usually, recognizing those digits as 504192 is easy, and that ease is deceptive. In each hemisphere of the human brain, there is a primary visual cortex called V1, which contains about 140 million neurons, with tens of billions of connections between them. For humans, recognizing the above number is based on learning from experiences.

504192

Neural networks approach the problem differently to recognize handwritten information than a human brain by taking many handwritten digits for training and then developing a system to learn from those examples.

Fortunately, there is extensive handwriting data available from MNIST. The database is widely used for training and testing in machine learning. The MNIST data consists of two parts. The first part contains 60,000 images that can be used for training data, which are handwriting samples scanned from 250 people.

The second portion of the dataset is 10,000 images to be used as test data. These are 28 by 28 greyscale images[37].

Chapter 4 Artificial Intelligence and Machine Learning

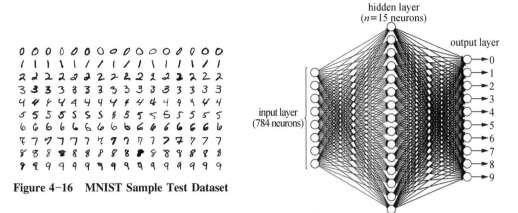

Figure 4-16 MNIST Sample Test Dataset

Figure 4-17 Neural Networks and Deep Learning Architecture

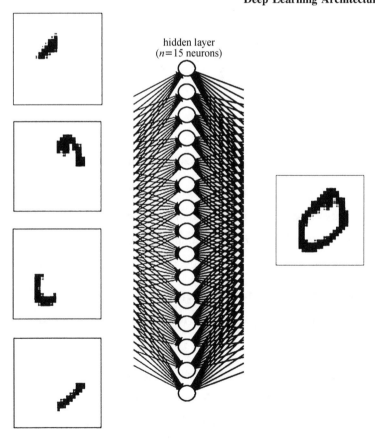

Figure 4-18 Neural Networks and Deep Learning-handwritten Digits Recognition

Architecture of Neural Network

The architectures of deep learning, like deep neural networks, deep reinforcement learning, deep reinforcement learning, recurrent neural networks, convolutional neural networks, and deep belief networks have been applied to many fields, including computer vision, natural language processing, speech recognition, machine translation, medical image analysis, bioinformatics, drug design, climate science, board game programs, where they have to provide results not only comparable to and in some cases surpassing the human expert performance[40].

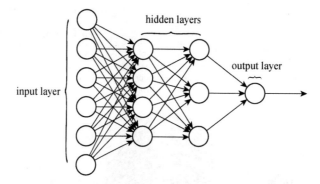

Figure 4-19 Layers in Neural Network Architecture

Feed forward neural network is an artificial neural network where the node connections do not form a cycle. They are biologically inspired algorithms with several neuron-like units arranged in layers. The units in neural networks are connected and are called nodes. Data enters the network at the input point and seeps through every layer before reaching the output.

We have been discussing neural networks where the output from one layer is used as input for the next layer. Such networks are called feedforward neural networks. Otherwise, we call it a recurrent neural network.

Deep neural networks (DNNs) are feed forward neural networks (FFNNs) where data moves from the input layer to the output layer only in the forward direction. The links between the layers are only in the forward direction, and they never touch a node again. Deep learning or deep neural networks (DNN) architecture consists of multiple layers, specifically hidden between the input and output layers. [41]

Chapter 4 Artificial Intelligence and Machine Learning

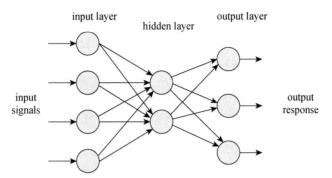

Figure 4-20 Feedforward Neural Network

Another class of deep neural networks is the convolutional neural network (CNN or ConvNet). CNNs are most commonly used in computer vision. By giving several images or videos from the physical world, using CNN, the system learns to extract the features of these inputs automatically to complete a specific task, e.g. face authentication, image classification, and semantic segmentation.

In contrast with fully connected layers in MLPs, in CNN models, there are one or multiple convolution layers extracting simple features from the input, which is done by executing convolution operations. Each layer of CNN is a set of nonlinear functions of weighted sums at different coordinates of spatially nearby subsets of outputs from the last layer, which allows the weights to be reused.

Recurrent neural networks (RNNs)

A recurrent neural network is a division of artificial neural networks where connections between nodes form a directed or undirected graph along a temporal sequence. This allows it to exhibit temporal dynamic behaviour. Derived from feedforward neural networks, RNNs can use their internal state (memory) to process variable-length sequences of inputs.

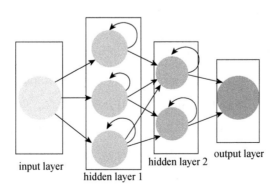

Figure 4-21 Recurrent Neural Network

Long short-term memory (LSTM) can learn long-term dependencies, making RNN smart at remembering what has happened in the past and finding patterns across time to make its next guesses make sense. LSTMs broke records for improved Machine Translation, Language Modelling and Multilingual Language Processing. Unlike standard feedforward neural networks, LSTM has feedback connections and can process not only single data points but also entire data sequences.

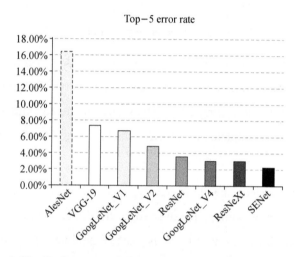

Figure 4-22 Performance of Current Popular Deep Neural Networks on ImageNet (human achieve an error rate of 5%)[42]

Deep Learning Frameworks

What is deep learning neural network?

Deep learning is used for more than just face recognition or voice recognition. Many more areas have benefited from deep learning and transformed our industries to an intellectual level at the beginning of the Industrial Revolution 4.

There are many deep learning frameworks, including:

TensorFlow is an open-source software library for machine learning and artificial intelligence. Developed by Google Brain, TensorFlow is, by far, one of the most used deep learning frameworks. TensorFlow provides a wide range of APIs (application programming languages), from pre-processing to data modelling. It is written in Python, C++ and

CUDA, and it runs on almost all platforms—Linux, Windows, macOS, iOS and Android. TensorFlow API is widely used in Python, entirely under stable releases.

Torch is an open-source machine learning library, a scientific computing framework, and a scripting language based on the Lua programming language. It provides a wide range of algorithms for deep learning, uses the scripting language LuaJIT, and has an underlying C implementation. It was created at IDIAP at EPFL.

Facebook's AI Research Lab develops PyTorch, another widely used deep learning framework mainly for its Python interface. PyTorch is built on top of the Torch library, and PyTorch has an active development community for computer vision and NLP to reinforcement learning techniques. Some milestones by PyTorch are Hugging Faces Transformers, PyTorch Lightening, Tesla Autopilot, Uber Pyro, and Catalyst.

KERAS is a cross-platform neural network library written in Python, developed by Francis Chollet. KERAS is a high-level API built on top of TensorFlow. KERAS is the most used deep learning framework in Kaggle, and KERAS best runs on GPUs and TPUs.

Apache Software Foundation develops Apache MXNet; MXNet is an open-source deep learning framework built for high scalability and support by various programming languages. MXNet is written in multiple languages—C++, Python, Java, Scala, Julia, R, JavaScript, Perl, Go and Wolfram Language. It is known for its fast model training. Apache MXNet supports deep learning models, such as convolutional neural networks (CNNs) and long short-term memory (LSTM).

CAFFE is a deep learning framework originally developed at the University of California, Berkeley, AI Research (BAIR). It is open source, under a BSD license. It is written in C++ with a Python interface. Caffe supports various image segmentation and classification architectures. Caffe is written in C++. Compatible with Linux, Windows, and macOS. It works on CPUs but has better performance with GPU acceleration. CAFFE is preferred for its speed and industry deployment. It can process up to 60 million images with NVIDIA GPU.

Project: Intelligent Waste Bin

Intelligent waste bin for smart cities

This is a research project conducted in our robotics and AI labs. The project aim was developing an intelligent bin that automatically opens relevant waste bin doors by detecting a recyclable object in a person's hand while walking toward the bin. In addition, each deposited trash bag was calculated by weight, type, time, battery status, and bin container status. The data is uploaded to a remote database for community bin management.

The project already went into production in 2021 and is currently operational in selected residential communities in Huai'an city.

Hardware
1—Jetson TX2
2—Arduino Mega Microcontroller
3—Arduino UNO Microcontroller
4—Actuator / Motor Drivers
5—Ultrasonic sensors
6—Weight sensors
7—Vertical door actuators
8—Infra-Red sensors
9—Electric door locks
10—PSU
11—Battery

Software
1—Yolo 2
2—MySQL
3—Django

Chapter 4 Artificial Intelligence and Machine Learning

AI Ready Devices

OAK-D Camera (OpenCV AI Kit)

Figure 4-23　OAK-D Camera

The OAK-D baseboard has three on-board cameras which implement stereo and RGB vision, piped directly into the DepthAI SoM for depth and AI processing. The data is then output to a host via USB 3.1 Gen1 (Type-C).

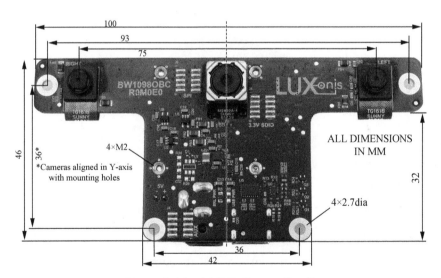

Figure 4-24　OAK-D Camera Details

Table 4-2 OAK-D Camera Specifications

Camera Specs	Colour camera	Stereo pair
Sensor	IMX378	OV9282
DFOV/HFOV/VFOV	81°/69°/55°	82°/72°/50°
Resolution	12 MP (4032 × 3040)	1 MP (1280 × 800)
Focus	Auto-Focus: 8 cm-∞	Fixed-Focus: 19.6 cm-∞
Max framerate	60 FPS	120 FPS
F-number	2.0	2.2
Lens size	1/2.3 inch	1/4 inch
Pixel size	1.55 μm × 1.55 μm	3 μm × 3 μm

This OAK camera has on-board **Myriad X** VPU. Main features:

Take your imaging, computer vision and machine intelligence applications into network edge devices with the newest Movidius family of vision processing units (VPUs) by Intel.

- **4 TOPS** of processing power (1.4 TOPS for AI)
- **Run any AI model**, even custom architecture/built ones—models need to be converted
- **Encoding**: H.264, H.265, MJPEG −4K/30FPS, 1080P/60FPS
- **Computer vision**: warp/dewarp, resize, crop via ImageManip node, edge detection, feature tracking. You can also run custom CV functions
- **Stereo depth** perception at with filtering, post-processing, RGB-depth alignment, and high configurability
- **Object tracking**: 2D and 3D tracking with ObjectTracker node

Chapter 4   Artificial Intelligence and Machine Learning

Project: Road Detection

Outdoor Robot—Road Detection

RIMA TX1 was a complete indoor robot build in our robotics lab based on Turtlebot 2. There were three main parts of design.
1—Body with raising neck, and Arm
2—Voice recognition and text to Speech
3—Chatbot development

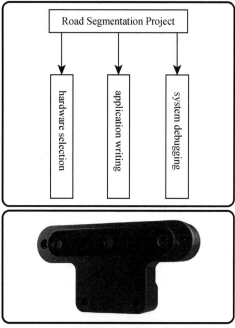

StudentProject: Outdoor Robot Road Detection

By Fei Hongyan

The project aim is to use OpenCV OAK-D camera for the fastest detection of road without using extra hardware.

The idea of the program is to download the road segmentation model from the OpenVINO Model Zoo and perform inference on the RGB camera, as well as the depth output. Think of the road as a segmentation network and classify each pixel into four categories: BG, road, curb, and marker. Take a shape blob with an image and format BGR, where B-batch size, C-number of channels, H-image height, and W-image width are input, output a blob whose format B, C, H, W. It can be treated as a four-channel feature map, where each channel is the probability of one of the following classes: BG, road, curb, marker.

The design framework of the program, first import the required library and define the pipeline, secondly set the camera (create the camera, select the required camera, and set the resolution of the camera), and finally, create the pipeline when the pipeline is defined, we can connect the device.

To create pipeline:

```
pm = PipelineManager( ) pm. createColorCam( previewSize = nn _ shape) nm = NNetManager( inputSize = nn_shape) pm. setNnManager( nm) pm. addNn( nm. createNN ( pm. pipeline, pm. nodes, blobconverter. from_zoo( name = 'road-segmentation-adas-0001', shaves =6)),
sync = True)
fps = FPSHandler( )
pv = PreviewManager( display = [ Previews. color. name] , fpsHandler = fps)
```

Define pipes, connect equipment

```
withdai. Device( pm. pipeline) as device:
nm. createQueues( device)
pv. createQueues( device)
```

Chapter 4　Artificial Intelligence and Machine Learning

Joint debugging

The joint debugging of road surface segmentation is a comprehensive test of software and hardware under the premise that the hardware circuit debugging is normal and the software system debugging is normal, mainly to verify whether the camera can shoot and meet the functional requirements. The specific joint debugging steps are as follows.

Step1: Install dependencies

```
Python3-m pip install -requirements. txt
```

The result of a successful installation is as follows:

Step2: Run the driver

```
Python3 main. py
```

The effect of running the rendering is as follows:

Human Intelligence vs. Robot Intelligence

Only two decades ago, the comparison between humans' and robots' intelligence was considered a joke. There was no doubt that robots could not do anything as humans do for a long time. There is still a vast gap between human and robot intelligence, but there are areas in which machine learning proved that it could do better than humans.

- Repetition
- Precision
- Immunity and hazard

We always think of AI's strength as convergent intelligence. In challenges that need high memory and processing power, they outperform a human at rules-based games, complex calculations, and data storage, such as chess and advanced math. The results of many competitions between humans and computers proved such superiority during the last decade.

We may argue that computers lack imagination and rule-breaking curiosity—that is, divergence.

While we (humans) have had these arguments for decades, artificial intelligence and machine learning are evolving much faster than humans in the last 4 billion years. The result, only during the last ten years, is self-driving cars, mobile robots, voice and face recognition, generative design, and robotic surgery.

This fast evolution of robots will cause a fast rise of problems in human-robot interaction. While robots more and more occupying our space, there will be a need for a human-robot protocol to:

1—Keep humans safe from trouble caused by robots;
2—Keep robots performing better and more productive.

How Robots See the World

Robots can see the world through digital cameras, radars, ultrasonic sonars, and infrared. There seems to be a lot of data that provides enough information about the scene.

Figure 4-25 An Image that is Generated by Robotic Eye

Chapter 4 Artificial Intelligence and Machine Learning

With our experience in a city, road, traffics, rules, and many more, we can interpret the above image much better than an onboard computer of a self-driving car.

When an autonomous car scans a person with its forward-facing radar, they show the same reflectivity as a soda can, explains SvenBeiker, executive director of the Centre for Automotive Research at Stanford University. "That tells us that radar is not the best instrument to detect people. The laser or especially camera is more suited to do that."

However, for us, it is not a matter of only a data set of similar objects and labelling. For instance, consider a glass or a cup. Despite its possible strange unexperienced shapes like a jug, or a little pot, we may correctly recognise it based on many other factors. For us, it is not only about the object only. We immediately consider the environment, where it is positioned, and how it is positioned. For example, if it is related to planting, we may consider it as a pot.

For a human, understanding an object is related to many other things (physical or non-physical). Here I use the word "thing" because it may be other objects, a way of using them in relation to an event, or perhaps a new social rule that shapes our way of seeing.

Perhaps each object can be referred to as a "dispositif", which not only explains a label but also a network of relationships with other dispositifs, making it easier for us (humans) to recognize objects.

Dispositive Networks and Understanding of the World

As much as we all are unique individuals in social aspects, cultural resources influence our model of thought and actions. This difference in the way of seeing the world resulted in Agamben's development of Foucault's use of the term "dispositif." Agamben (2009) claims a dispositif is a heterogeneous set that is partitioned into two large groups or classes, living beings and everything else, which originated at some point in human history, with the specific purpose of defining or shaping our behaviour. Examples of such "dispositifs" include "communication," "technological devices," "law," and "education". However, various activities common to all humans are employed differently from one social group to another and can produce different ways of thinking and behaving. It is then that dispositifs can become the captor of our behaviour. An example is the mobile phone was developed to address a problem and a need in society but later became the captor of our way of thinking and behaving in relation to our social communication. The same dispositif does not ultimately shape people's behaviour in the same way everywhere else[43].

Then our understanding of the world around us is based on interconnected dispositifs. A multi-layer network connects to items from different times, fields, and nature in our mind.

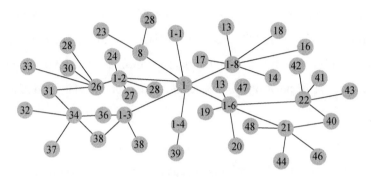

Figure 4-26　Dispositifs Network

Dispositifs are educating our way of learning because of their power of network structure. That means each dispositif is in a influential relation with other dispositifs. Therefore, our way of thinking is developed based on a network of dispositifs, which may be common to all social groups but may be employed in a different way. This difference creates a different way of seeing and acting.

Human-Robot Interaction

Safety must be a primary focus of self-driving cars' Human-Robot Interaction (HRI). Currently, road safety is one of the growing concerns in the global dimension. According to a global status report on the road safety performed by the World Health Organization (WHO) in 2015, there were over 1.25 million traffic-related accidents annually, leaving tragic consequences[44]. Accelerating technological developments in robotics during the last decade encouraged researchers, developers, and the industry to invest in deploying robotics

Chapter 4 Artificial Intelligence and Machine Learning

solutions to traffic-related accidents.

For engineers, mathematicians, AI architectures, and computer programmers, the question is simple: "How to detect a pedestrian human?" or "How to avoid crashing into another car?"

They try to solve these problems by using different sensors, algorithms or even faster detection. That is, in our mind, a computer is designed and capable for:

Accuracy, high memory, speed, …

However, for social scientists, the social psychologist problem still needs to be solved by even precise detection of a pedestrian in a short time. There still needs to be a solution to understanding intentions. Within the same social group, we are familiar with other drivers' or pedestrians' way of thinking. Therefore, it is relatively easy to detect a human actor's intention on the road as a pedestrian or a driver. This familiarity with the way of thinking of others is the result of sharing a similar network of dispositive. However, the same robot or autonomous car with the same precise algorithm may not be able to do the same in another society.

In one incident,

"Less than two hours after departing, Las Vegas' new, self-driving shuttle crashed with a delivery truck without any injuries. Jenny Wang, one of the passengers on the shuttle at the time of the crash, told a local news station that the shuttle just stayed still, and they knew the truck was going to hit them, and then it hit them. The shuttle did what it was expected to do. That was a full stop to avoid an accident; however, the delivery truck did not stop, grazing the shuttle's front fender"[43].

In another incident, it has been reported:

"In another instance, just less than a mile from start-up, a self-driving test car, from Ford-backed start-up, Argo, containing four passengers, was hit by a van jumping a red light in Pittsburgh. Four occupant passengers of the self-driving car were injured and transported to the hospital."[43].

Most accidents involving self-driving cars were mainly the result of the self-driving cars' inability to predict human intentions. By using their sensors and intelligent algorithms, self-driving cars can detect objects and movements and choose the appropriate and desired action. Driving between humans for a self-driving car (autonomous car) is best to act like human drivers for the safety of both. Then acting like a human would require understanding a human's way of thinking.

Alternatively, our roads and traffic environments must be robot-friendly, and human drivers act like self-driving cars with a rule-based way of thinking.

Exercises and Projects

Changes in the field of robotics are speedy. Therefore, to provide more UpToDate, more up-to-date problem-solving techniques and innovative projects, this part will be available from the author's website at www.roboticacamp.com/robotics_and_ai.

References

[1] CSC 297 Robot Construction: Electronics [EB/OL]. (2022-03-13) [2022-09-01]. https://www.cs.rochester.edu/u/nelson/courses/csc_robocon/robot_manual/electronics.html.

[2] Robert Balmer, William Keat. Exploring Engineering[M]. 5th ed. America: Elsevier, 2022.

[3] Binary Logic, Number systems, Gates[EB/OL]. [2022-03-13]. https://ieda.ust.hk/dfaculty/ajay/courses/alp/ieem110/lecs/numsys/numsys.html.

[4] UNO R3 | Arduino Documentation[EB/OL]. [2023-01-29]. https://docs.arduino.cc/hardware/unorev3.

[5] wiki_EN_content_page-DFRobot[EB/OL]. [2022-01-29]. https://wiki.dfrobot.com/.

[6] Nano 33 BLE Sense | Arduino Documentation[EB/OL]. [2023-01-29]. https://docs.arduino.cc/hardware/nano-33-ble-sense.

[7] Pocuter One[EB/OL]. [2023-01-29]. https://www.pocuter.com/pocuter-one.

[8] R. Watson, "How to Program the STM32 'Blue Pill' with Arduino IDE | Arduino," *Maker Pro* [EB/OL]. (2019-07-06) [2023-09-01]. https://maker.pro/arduino/tutorial/howto-program-the-stm32-blue-pill-with-arduino-ide.

[9] Raspberry Pi Documentation-Raspberry Pi Pico [EB/OL]. [2022-03-19]. https://www.raspberrypi.com/documentation/microcontrollers/raspberry-pipico.html.

[10] Teensy. *Cool Components*[EB/OL]. [2023-01-29]. https://coolcomponents.co.uk/collections/teensy.

[11] MicroPython-Python for microcontrollers[EB/OL]. [2023-01-29]. http://micropython.

org/.

[12] WiFi Kit 32 (Phaseout). *Heltec Automation*[EB/OL]. (2019-04-23) [2023-09-01]. https://heltec.org/project/wifi-kit-32/.

[13] OpenCR1.0 - ROBOTIS[EB/OL]. [2023-01-29]. https://www.robotis.us/opencr1-0/.

[14] I2C - learn.sparkfun.com[EB/OL]. [2023-01-29]. https://learn.sparkfun.com/tutorials/i2c/all.

[15] How To Use Arduino's Analog and Digital Input/Output (I/O)-Projects[EB/OL]. [2022-05-19]. https://www.allaboutcircuits.com/projects/using-the-arduinosanalog-io/.

[16] Jetson TX2 Module. *NVIDIA Developer* [EB/OL]. (2017-05-01) [2023-09-01]. https://developer.nvidia.com/embedded/jetson-tx2.

[17] Handbook of Modern Sensors: Physics, Designs, and Applications by Jacob Fraden-PDF Drive[EB/OL]. [2022-03-20]. https://www.pdfdrive.com/handbook-ofmodern-sensors-physics-designs-and-applications-e8368354.html.

[18] LDR Sensor Module Interface With Arduino[EB/OL]. [2023-01-31]. https://www.instructables.com/LDR-Sensor-Module-Users-Manual-V10/.

[19] Thermistor. *Wikipedia*[EB/OL]. [2022-05-17]. https://en.wikipedia.org/w/index.php?title=Thermistor&oldid=1077657443.

[20] openQCM the Temperature Sensor Using a Thermistor with Arduino | Quartz Crystal Microbalance QCM-D with Dissipation Monitoring: the first scientific QCM entirely Open Source [EB/OL]. (2015-03-17) [2022-02-01]. https://openqcm.com/openqcm-temperature-sensor-using-a-thermistor-witharduino.html.

[21] 5 Types of Pro imity Sensors (Application and Advantages) | Linquip[EB/OL]. (2021-04-01) [2023-02-01]. https://www.linquip.com/blog/types-of-proximity-sensors/.

[22] PID controller. *Wikipedia*[EB/OL]. [2022-03-12]. https://en.wikipedia.org/w/index.php?title=PID_controller&oldid=1076655335.

[23] PID for Dummies-Control Solutions[EB/OL]. [2022-03-19]. https://www.csimn.com/CSI_pages/PIDforDummies.html.

References

[24] Blink Without Delay | Arduino Documentation [EB/OL]. [2022-03-21]. https://docs.arduino.cc/built-in-examples/digital/BlinkWithoutDelay.

[25] A. Gupta, What is OpenCV and Why is it so popular? *Analytics Vidhya* [EB/OL]. (2019-07-24) [2023-09-01]. https://medium.com/analytics-vidhya/what-and-whyopencv-3b807ade73a0.

[26] ROS for Beginners: What is ROS?. *The Construct* [EB/OL]. (2019-09-20) [2022-03-22]. https://www.theconstructsim.com/what-is-ros/.

[27] R. A. TorricoBuild a—DOF Robotic Arm (Part 2) [EB/OL]. [2020-02-01]. https://circuitcellar.com/research-design-hub/build-a-4-dof-robotic-arm-part2/.

[28] turtlebot/Tutorials/indigo/Network Configuration—ROS Wiki [EB/OL]. [2022-04-14]. http://wiki.ros.org/turtlebot/Tutorials/indigo/Network%20Configuration.

[29] Ridgeback—Omnidirectional mobile manipulation robot. *Clearpath Robotics* [EB/OL]. [2022-04-19]. https://clearpathrobotics.com/ridgeback-indoor-robot-platform/.

[30] NAO the humanoid and programmable robot | SoftBank Robotics [EB/OL]. [2022-04-15]. https://www.softbankrobotics.com/emea/en/nao.

[31] Artificial Intelligence—Overview [EB/OL]. [2023-09-01]. https://www.tutorialspoint.com/artificial_intelligence/artificial_intelligence_overview.html.

[32] Neural network," *Wikipedia* [EB/OL]. [2023-09-01]. https://en.wikipedia.org/w/index.php?title=Neural_network&oldid=1077721872.

[33] U. of C.-S. Diego. Why are neuron a ons long and spindly? Study shows they're optimi ing signaling efficiency [EB/OL]. [2022-04-16]. https://medicalxpress.com/news/2018-07-neuron-axons-spindly-theyreoptimizing.html.

[34] Single Layer Perceptron in TensorFlow-Javatpoint [EB/OL]. [2022-04-16]. https://www.javatpoint.com/single-layer-perceptron-in-tensorflow.

[35] Tensor Flow-Multi-Layer Perceptron Learning [EB/OL]. [2022-04-16]. https://www.tutorialspoint.com/tensorflow/tensorflow_multi_layer_perceptron_learning.html.

[36] Activation function. *Wikipedia* [EB/OL]. (2022-03-09) [2022-04-17]. https://en.wikipedia.org/w/index.php?title=Activation_function&oldid=1076034609.

[37] M A Nielsen. Neural Networks and Deep Learning[EB/OL]. [2022-04-17]. http://neuralnetworksanddeeplearning.com

[38] Gradient Descent-Gradient descent-Product Manager's Artificial Intelligence Learning Library [EB/OL]. [2022-04-17]. https://easyai.tech/en/ai-definition/gradient-descent/.

[39] Oticon BrainHearingTM [EB/OL]. [2022-04-17]. https://www.oticon.com/yourhearing/hearing-health/brainhearing-technology.

[40] D Ciregan, U. Meier, and J. Schmidhuber, "Multi-column deep neural networks for image classification," in 2012 *IEEE Conference on Computer Vision and Pattern Recognition*, Jun. 2012, pp. 3642-3649. doi: 10.1109/CVPR.2012.6248110.

[41] S LABS. Understanding Deep Learning: DNN, RNN, LSTM, CNN and R-CNN[EB/OL]. (2019-03-21)[2022-04-17]. https://medium.com/@sprhlabs/understanding-deep-learning-dnn-rnn-lstmcnn-and-r-cnn-6602ed94dbff.

[42] W. Du et al. Review on the Applications of Deep Learning in the Analysis of Gastrointestinal Endoscopy Images[J]. *IEEE Access*, 2019, vol. 7, 142053-142069, doi: 10.1109/ACCESS.2019.2944676.

[43] A A Mokhtarzadeh and Z. J. Yangqing. Human-Robot Interaction and Self-Driving Cars Safety Integration of Dispositif Networks[J]. IEEE International Conference on Intelligence and Safety for Robotics (ISR), Shenyang, China, 2018, 494-499, doi: 10.1109/IISR.2018.8535696.

[44] World Health Organization. Global status report on road safety 2015[M]. World Health Organization, 2015.

[45] Arduino Uno Rev3 SMD. *Arduino Official Store*[EB/OL]. [2023-01-29]. https://store.arduino.cc/products/arduino-uno-rev3-smd.